E.W.R. STEACIE AND SCIENCE IN CANADA

E.W.R. Steacie,
scientist, teacher, and leader, 1961.
Photo by Malak, Ottawa

M. CHRISTINE KING

E.W.R. STEACIE
and Science in Canada

UNIVERSITY OF TORONTO PRESS
Toronto Buffalo London

© University of Toronto Press 1989
Toronto Buffalo London
Printed in Canada

ISBN 0-8020-2667-2

Printed on acid-free paper

Canadian Cataloguing in Publication Data

King, M. Christine, 1947–1987
E.W.R. Steacie and science in Canada

Bibliography: p.
Includes index.
ISBN 0-8020-2667-2

1. Steacie, Edgar William Richard, 1900–1962.
2. National Research Council of Canada – History.
3. Science – Canada – History. 4. Chemists –
Canada – Bibliography. I. Title.

QD22.S85K56 1989   540'.92'4   c89-093144-5

Contents

FOREWORD vii
PREFACE ix

1 Early Days and McGill 3

2 A Taste of Europe and the Years After 18

3 A Chemist Finds His Calling 34

4 The Making of an Institution: The National Research Council of Canada 45

5 Progress through Problems 60

6 Lessons in Scientific Diplomacy 77

7 Apprentice in Statesmanship 94

8 Leader in an Age of Certainty 116

9 The Politics of Science 129

10 Science and Government: The Heeney Report 147

11 Years of Fulfilment 165

12 Final Days and Unfinished Business 186

13 Epilogue 195

APPENDIX: E.W.R. STEACIE'S SCIENTIFIC PUBLICATIONS 201
NOTES 215
BIBLIOGRAPHY 235
INDEX 239

Foreword

Shortly after completing the manuscript for this book, the author, Dr Christine King, died in an automobile accident in England.

Early in life, Dr King was attracted to chemistry, an attraction which culminated in a doctoral degree in chemical physics. Increasingly, however, she became interested in the role which science played in the history of different countries. Further work along these lines led to a second doctoral degree in the history and philosophy of science and to the decision to make this field her vocation.

Within the broad field of the history of science, and no doubt influenced by her early training, Dr King focused on the development of chemical kinetics. Her research on this topic brought her to Canada where, in association with Dr K.J. Laidler at the University of Ottawa, she continued her historical work. While there she became increasingly interested in the history of Canadian science and discovered that both of these subjects were strongly combined in the life and work of E.W.R. Steacie. It was at this point that the National Research Council of Canada invited her to undertake this biography.

At the time of Dr King's death the manuscript had been accepted for publication but the work of editing had barely begun. With the sympathetic co-operation of the University of Toronto Press it was agreed that the arrangements for publication should continue and I was asked to serve as adviser. In these circumstances, the judgment and skills of the editor, Diane Mew, were called on to an exceptional degree and her contribution to the finished work is invaluable.

The book which has emerged is more than a simple biography. It deals, of course, with the life of E.W.R. Steacie and the part he played in the development of Canada's National Research Council and of science in

Canada generally. But it also provides much information on the relations between the NRC and the Canadian government, universities, and industry, and on scientific links with Britain and the United States, in the important and formative decades during and following the Second World War.

Sadly, following her untimely death, the book also stands as a memorial to Christine King and a measure of our loss at her passing.

A.W. Tickner

Preface

Edgar William Richard Steacie was born on Christmas Day in the year 1900. Europe then dominated a world which, though poised on the edge of a new century, was still tied to the past. As if to make up for lost time, the pace of change in the coming decades would accelerate increasingly, and those changes would be brought about largely by advances in science and technology. At the century's beginning, the atom was hardly understood and aeroplanes were unheard of; by 1962, when Steacie died, atoms had long since been taken apart and put together again, and man had entered space.

Canada – 'a small country, battling against vast geography' – was in the throes of making her own history. The major scientific role in this history would be played by the National Research Council of Canada, created in 1916 and still at a stage of pliable youth when Steacie joined it in 1939. Eventually, in a period of leadership spanning just ten years, Steacie would take the NRC and mould it into an international organization, foremost in several areas of scientific endeavours, including his own field of gas phase kinetics. Wherever Canadian scientific effort stood on the map at the beginning of Steacie's era, it occupied a far larger area at the end.

When this end came, Steacie was standing at the pinnacle of his career and personal power, admired by colleagues at home and abroad, respected by scientific leaders throughout the world. He had been honoured by the premier scientific bodies of two nations diametrically opposed in their national philosophy: he had lectured on chemistry to both the American Academy of Sciences and the Soviet Academy of Sciences and voiced his policies on the national role of science to each of them. He was president of the world's leading scientific congress, the

International Council of Scientific Unions. Few distinctions remained to be had, although much work, he knew, still awaited. It was then that Steacie experienced the obstructing, unforgiving nature of time.

Less than a decade after his death, Steacie would be the subject of a national controversy, which at its height threatened to fragment the very ideals to which he had devoted the greater part of his working life and energies. Steacie, however, had lived and presided over quite a different era; life in the Ottawa of the fifties was uniquely Canadian. True, science was in ascendancy everywhere, but Canada was poised for a simultaneous period of unprecedented economic prosperity and a newly won spirit of national confidence.

It was a time and place eminently suited to Steacie, a natural leader, with genuine abilities in the unlikely combination of scientific research and administration. To these were added an incisiveness that invited loyalty, and an innate charm that quickly captivated those he came in contact with, both within the institution he would lead and beyond. In a career spanning several decades Steacie left almost no correspondence, diaries, or notes of a personal nature. This reticence tells us much about the man but is of little comfort in recreating the everyday nuances through which a life is lived. What remains is a collection of official correspondence from his office as president of the National Research Council and a large number of public speeches addressing the issues of his day.

A few notes of explanation are thus called for. The interval between 1939 and 1950 was formative in the emergence of the National Research Council as the country's pre-eminent scientific institution. While Steacie himself neglected to keep any record, it is fortunate that C.J. Mackenzie, leader of the NRC during this period, left a series of diaries and a collection of correspondence which permit an insight into developments in this era. Over the years, the respective roles played by Mackenzie and his younger colleague Steacie would evolve into a remarkable symbiotic relationship which would affect the subsequent development of science in Canada. Yet, though united by a common purpose, the two men were distinctly different in character. Mackenzie's strength during the difficult times of war derived from an unusual degree of calm common sense rather than as a result of intellectual planning. Steacie's success, by contrast, depended on an exceptional ability for clear thinking and decisive action in the midst of vast complexities. Each of these qualities was essential to the era of their leadership. Indeed, if fate had reversed Steacie's and Mackenzie's positions, it is doubtful whether there would be much to record about either period of NRC history. Few scientific institutions have achieved the international status won by the NRC during Steacie's leadership so quickly yet built upon such sound foundations. Without describing Mackenzie's role in building up this strong base at the NRC in the preceding years, Steacie's rapid success would appear illogical and unbelievable. For these reasons, in this story Mackenzie's role has been given considerable emphasis, for it explains much of what happened afterward.

The subject of chemical kinetics is not treated here in any great depth, although an overview of Steacie's achievements as a scientist is included. The rapid advance of chemical data has far superseded the extent of knowledge in Steacie's day, whereas the impact of Steacie's personality and his contributions to policies affecting science have proved extraordinarily lasting.

It is one of the ironies of history that those who sought to keep his memory alive in the mid-sixties envisaged quite the opposite. Leo Marion, opening the Steacie Building for Chemistry at Carleton University in 1965, observed: 'The picture of the man will gradually fade in the memories of people, and what will remain will be whatever part of his scientific work that will prove eventually to have contributed most to scientific knowledge.' Over twenty-five years have passed since Steacie's death, yet the image of the man has remained sharply etched in the minds of many people in several countries. While those who knew him and lived through that same era still remembered, it seems important to preserve some facets of this chapter in a continuing history – a chapter in which people and events came together in a way unlikely to be repeated. But the real tribute to Steacie must reside in the life and work of the generations that he influenced and who endeavour to impart the same inspiration to others, as is evident in different parts of Canada, Britain, and the United States today.

I am greatly indebted to a large number of Steacie's former students, colleagues, friends, and NRC post-doctoral fellows who travelled with him at various stages of events covered in this book and responded to my request for information. It is only the contributions and recollections of these people that have enabled me to flesh out the story of a man of whom so few personal records remain. Those providing information included: Dr P.J. Agius, Dr M. Baker, Prof. F.E. Blacet, Prof. J.G. Calvert, Dr W.H. Cook, Prof. I.M. Cowan, Dr R. Cvetanovic, Lord Dainton, Dr J. Dewar, Dr P. Gishler, Dr N.T. Gridgeman, Dr S.S. Grimley, Prof. H.E. Gunning, Dr J. Harrison, Dr G. Herzberg, Dr L. Howlett, Prof. K.J. Ivin, Dr L. Kerwin,

Dr K.O. Kutschke, General G. Laclavere, Prof. K.J. Laidler, Dr R. Legget, Prof. D.J. Leroy, Dr F.P. Lossing, Dr Carol Maass, Dr T.W. Martin, Dr C.R. Masson, George McColm, Dr K.A. McLauchlin, Dr P.D. McTaggart-Cowan, Sir Harry W. Melville, Dr and Mrs Mercer, Dr W.E.K. Middleton, Prof. R.V.V. Nicholls, Prof. C. Ouellette, Dr N.W.F. Phillips, Prof. J.N. Pitts, Prof. J. Polanyi, Sir George Porter, Dr I.E. Puddington, Prof. B.S. Rabinovitch, Prof. H. Schiff, Dr W.G. Schneider, Prof. G.M Shrum, Dr O.M. Solandt, Dr B.C. Spall, Prof. J.W.T. Spinks, Prof. D.S. Tarbell, Mel Thistle, Prof. H.G. Thode, Prof. A.F. Trotman-Dickenson, Prof. E. Whittle, Prof. L. Yaffe, and Mrs E. Zimmerman.

My special thanks are due to Dr A.W. Tickner of the NRC for much assistance and many enjoyable discussions and to Professor J.A. Morrison of McMaster University for his kind interest throughout this study; to Denice Willis, Dene McColm, Denis Pinard, and to the staff of the Canada Institute for Scientific and Technical Information for their tireless help in tracing sources; to Mrs H. Cuccaro for her skilled typing of the script. My sincere thanks also go to Dr D.J. Ball and Karen King for helpful criticisms.

Various institutions have kindly provided information, in particular, the Royal Society of London, the Royal Society of Canada, McGill University (Archives), University of Toronto (Archives), and Bibliotek und Archiv zur Geschichte der Max-Planck-Gesellschaft, Berlin.

To the staff, academic and secretarial, of the history department of Carleton University, who so generously permitted me to share their facilities during the course of this work, my thanks are barely adequate; the campus view of the Rideau River they provided will not be forgotten.

I am particularly grateful to Mrs Steacie and her family for their patience in answering countless questions and permitting me to intrude so frequently and persistently into their lives during my research. Any errors or omissions are of course entirely mine.

M. Christine King
Ottawa and London

Baby Ned, 1901.
Courtesy Mrs Steacie

The young Ned Steacie at the Royal Military College, 1920.
Courtesy Mrs Steacie

Graduating from McGill University in 1923
with a first-class honours degree and
the British Association Medal.
Courtesy Mrs Steacie

The new president of NRC, 1952.
Courtesy Mrs Steacie

The Steacie cottage at Grand Lake, in the Gatineau Hills,
over the years a haven of quiet from increasing responsibility at
the National Research Council.
Courtesy Mrs Steacie

As head of the chemistry division, NRC,
Steacie at work in his lab, 1948.
Courtesy PAC. PA-145317

Steacie, now president of NRC,
in front of the Applied Chemistry Building of NRC.
Courtesy NRC

Steacie with Ira Puddington,
head of the applied chemistry division.
Courtesy NRC

At the meeting of the British Association in India, 1959.
Front row, l to r., Sir Harry Melville, Prince Philip, Steacie,
and Professor M.S. Thaker.
Courtesy Mrs Steacie

Steacie and Dr Wilder Penfield, foreign members of the USSR
Academy of Sciences, receive certificates at the Soviet Embassy in Ottawa, 1958.
Courtesy NRC

Dr F.S. Diatchkovsky, the first exchange scientist
from the Soviet Union to Canada,
with Steacie at the NRC lab in Ottawa, 1960.
Courtesy NRC

E.W.R. STEACIE AND SCIENCE IN CANADA

1
Early Days and McGill

Steacie's ancestors made their home in Ireland. The break came late in the nineteenth century when Richard Steacie, together with his mother and sister, emigrated from Ballinasloe in County Galway to Canada and settled in Montreal. Richard carved a military career for himself and held the rank of captain by the time he married Alice Kate McWood. The McWoods were of English stock and had established themselves in a fashionable part of Montreal, on Dorchester Street West. Alice was the youngest of a large family which doted on her, with a devotion that would last into adult life. After she married Richard and set up a home of her own, Alice found, to her surprise, that running a household was mostly tedium. When baby Ned arrived on Christmas Day, 1900, she decided it was time to move the whole family back to her childhood haven. As for the baby, her own family's devotion quickly encompassed this Christmas addition also who, if photographs are to be trusted, was alert and co-operative. Baby Ned thus passed into the care of a grandmother, a fond Aunt Emma, and the servants.

Being an only child in a household of adults was probably not that much fun. The large house was sombre and hushed, not allowing much light or joy for a growing boy. The household was ruled over by a stern Victorian grandfather with whom Ned rarely came into contact, and a somewhat distant mother whose affections were not perhaps demonstrated in the way a child would understand. Although his father was present, it was a quiet, unobtrusive presence. For some reason, his parents decided not to send Ned to school when he had reached the usual age. From time to time, other children were brought in to play and in later years there would be visits to cousins, but it was a somewhat lonely childhood. When he was a little older his frequent companion was Dandy, a fine retriever;

Steacie retained an affection for Dandy all through life and as an adult his own family would always share their home with a pet. Neighbours might have observed the young Steacie sitting alone on the front steps of the house thinking the thoughts of his small world. There was probably little laughter – all the more surprising to discover that the man he became 'loved to laugh.'[1]

Finally, at the age of eight, Ned was sent to King's School in Westmount. Whether he enjoyed these first encounters with the outside world is hard to say, but when Alice developed tuberculosis a few years later Ned was sent even farther away to board at Bishop's College School in Lennoxville. The years of schooling ended at Westmount High, which he attended from 1912 until 1917. The young Steacie was an excellent student and left school with both the Commissioner's Leaving Scholarship and the Gold Medal. Although no special interest had emerged to guide his immediate direction, his attention was already focusing, if a little fuzzily, on chemistry, to which his teacher, William Mussels, had guided him. In years to come Steacie, like so many other scientists, would recall with gratitude the devotion of a special teacher in developing his curiosity.

But another event would put the pursuit of science, temporarily at least, out of mind. In 1914, shortly after the outbreak of war in Europe, Captain Steacie was posted overseas. Bidding goodbye to his father at the station was also the end of a chapter of childhood for Ned; his father gave him the keys to their home and bade him take care of his invalid mother. Now more than ever, the young Steacie found himself the nucleus of his mother's life, though this did not always mean the centre. Adjusting herself to the life of an invalid while her son was still a boy left her unable to encourage him in his academic pursuits, even if she had so wished. Then again, in keeping with the times, Alice was strongly drawn to religious practices; every evening, mother and son, then aged fourteen, withdrew to pray fervently and with many tears for the safe return of the captain. It was not a childhood that one could look back upon with much nostalgia. He would recall later that it puzzled him as a child not to be allowed to play with the other children and get thoroughly dirty as they did. In retrospect, his mother's anxieties that he should not succomb to TB had good cause; two of her sisters died of it, as had Ned's paternal grandmother.

The young Steacie's world would have been somewhat conservative and almost certainly Conservative. Years later, when her son was president of the National Research Council, Alice's natural pride in his achievements was mixed with abhorrence that he had fallen to 'working for a Liberal government.' Intellectually, there were the usual kinds of literature lying around the house. As a child, Steacie read avidly and he retained many of his childhood books all his life; Kipling, Dickens, and Thackeray were favourites, but he read many others, especially when he went to live for a time in the house of his uncle, General C.A. Smart. This was a splendid house with an equally grand library. But clearly there was little scientific influence of either an amateur or a professional nature; for one destined to become the leader of science in Canada, Steacie's childhood would appear somewhat unpromising. For young Ned, no intellectually minded parent, no central figure qualified to guide an impressionable mind already showing promise, emerged on the domestic scene. The natural curiosity of childhood largely spent alone took on special significance. Steacie remained essentially a very private man, although he also developed a generous streak of extrovertism.

Captain Steacie died in action at the second battle of Ypres in April 1915. Shortly after, Steacie and his mother moved to a small apartment on Grosvenor Avenue, there to remain, except for a brief period, until Ned's marriage. At the age of sixteen he enrolled in a science course at McGill University, but as the war progressed, his attention drifted from science to the thought of a military career. The sudden loss of a father whom he had loved dearly would have intensified the desire to follow in the army tradition. General C.A. Smart, his mother's brother-in-law, was even then overseas with the Canadian forces. Steacie entered the Royal Military College of Canada (RMC) in August 1919, with the distinction of having received an Order of Merit and attained the highest marks in the entrance exam.

That summer Steacie had his first taste of genuine freedom from the binds of formal Westmount life. His cousin, Everett Holmes, some years older than Ned, had returned from the war, suffering from exposure to gas. The family agreed that the two boys would benefit from an excursion into nature. This was to be a pioneering adventure and preparations were truly basic, in keeping with that great tradition. The boys acquired a simple canoe, filled it with a side of bacon, some oatmeal, flour, and lard, took a rifle as precaution and a hunting knife, and launched their expedition from Ste Anne's. Their destination was James Bay, a round trip of well over a thousand miles, even as the crow flies, and much more by water. It was genuine wilderness country and the two intrepid explorers met many a hazard along the way; often at night they slept on rocks near the water's edge and found themselves being investigated by some curious moose. There was a forest fire and they learnt quickly how to keep out of the way of the raging flames. They met no other signs of human intrusion as they progressed through the forest, apart from a few Indians who proved friendly. They shot a deer to replenish their dwindling food supply. This transpired to be a much bloodier experience than either had bargained for and most certainly put the pair off hunting and the taste of venison for life. The journey there and back served its purpose, helping to fade the memories of the war. Altogether it was an expedition which both would long remember, a high point of their youth, when summers seemed very long. Years afterward Steacie could not help recalling with a grin that if only his mother, who perpetually worried over her only son, could have pictured some of the more exciting moments of the trip, it would never have got off the ground and into water; and that would have been an opportunity lost.

Canada's Royal Military College at Kingston, Ontario, had been founded in 1876 and was modelled largely on the u.s. Military Academy at West Point. It aimed to provide a complete education in all branches of military tactics, fortification, and general scientific knowledge necessary for its officers. But the college intended its officers to be professional engineers at the same time as they would be qualified for 'cavalry, artillery and infantry.'[2]

The compulsory preliminary examination was comprehensive, covering mathematics, grammar and composition, geography, history, French, German, Latin, with further optional examination in algebra, Euclidean geometry, trigonometry, English literature, and additional British and Canadian history. What the RMC intended to instill in its young recruits was 'the spirit of service, of obedience, of persistence, of loyalty and of single minded devotion to a military purpose.' Education at the college thus provided a concentrated period of living according to rule and an ability at the end of it all, of resolution amid 'seeming chaos and horror.'[3]

The college was still in the throes of restoration and revival after the turmoil of the First World War when Steacie, along with 102 other candidates, sat the qualifying exam. Only thirty-one obtained the required 60 per cent qualifying marks in all the subjects listed, so those with only 50 per cent were also allowed through. This low success rate so alarmed the authorities that a review of the high entrance standard was called for and the number of subjects required reduced. Whatever the consequences of this on new recruits, it did not affect Steacie. At the end of the summer term a year later, Steacie outclassed sixty-one others to place first in mathematics, mechanics, geometry, and engineering, with a third in French, field sketching, and map reading.

Intellectual abilities notwithstanding, Steacie soon realized that loyalty to family tradition was not quite sufficient. The military life was proving not much to his liking. This was not due to any innate lack of discipline or even some hidden streak of indolence. Far from it. Steacie was to find working a fourteen-hour stint a normal pace in adult life and considered it all 'fun.' It was more that, like most independent spirits, he did not take well to being told what to do, as stipulated by a series of preset rules and regulations. He was not exactly a rebel; but all that drilling was tedious when he would far rather be occupied with some more intellectual exercise. Steacie admitted he had made an error, obtained his discharge from the RMC almost exactly a year after entering, and re-enrolled at McGill, this time in the practical area of chemical engineering. The RMC would not fade entirely from his life; and in the meantime he had made the acquaintance of Hartley Zimmerman, who would remain a lifelong friend. His discharge certificate from the Canadian Expeditionary Force cited: 'rank Corporal; age 19 years, 3 months; height 5' 8¾"; complexion fair; eyes blue; hair brown.'[4] To which an observer might have added: disposition cheerful; ability very considerable; future outlook exceptionally bright. Armed with these qualifications and gifts, the young Steacie began his adult training in earnest.

McGill University, in the opening decade of this century, was distinguished, among other things, by the presence of Ernest Rutherford, appointed MacDonald Professor of Physics in 1898 at the age of twenty-seven, and an even younger Frederick Soddy who joined the chemistry department, fresh from Oxford, in 1900. Together, they would make science history with studies on natural transmutation of elements and the radioactive properties of disintegrating atoms. In 1907 Rutherford left McGill to become director of physics at Manchester, won the Nobel Prize for chemistry a year later, and leapt from strength to strength at the Cavendish Laboratory in Cambridge. Soddy had returned to England in 1903 at which time the university had not yet awarded its first doctoral degree. By 1907, however, seven PH D students were registered; of these only one was to achieve the degree. Annie Louise MacLeod was awarded the first PH D in chemistry in 1910 on the strength of experimental work she had carried out at Bryn Mawr under the guidance of the distinguished American chemist E.P. Kohler. The second PH D in chemistry was awarded two years later for experimental work also carried out abroad, at the University of Berlin, under the illustrious Emil Fischer.[5] McGill's home-produced doctoral candidates in chemistry took off after the addition of Otto Maass to the faculty staff. In Steacie's opinion,

Maass essentially established the first real graduate school of science in Canada.

Of German-Swiss descent, Maass was born in New York City in 1890, and had the advantage of having the mathematician Julius Plücker as a great-uncle. After graduating from McGill, Maass had travelled to Berlin in 1913 to work under Walter Nernst. In those halcyon days of German scientific supremacy he shared laboratory space with future notables such as Arthur Eddington and Frederick Lindemann, whom Maass described years later as 'a most unpopular man in regards to his personality ... a vegetarian who ate only spinach.'[6] Fluent in German, Maass revelled in the exciting scientific atmosphere that enveloped the University of Berlin in that era and no doubt garnered there the ability to construct simple but highly effective apparatus, an art at which Nernst and his school excelled. It was to be a short experience, abruptly ended by the outbreak of war. By the autumn of 1914, Nernst had advised Maass to make himself scarce. When challenged at the German-Austrian border by guards, Maass bit his cheek, coughed up some blood and feigned the dreaded tuberculosis, and was quickly allowed through. From there he passed through Italy and Switzerland, finally returning to McGill, where he became a lecturer in physical chemistry. In 1918 Maass was given leave of absence to continue his interrupted studies at Harvard with Theodore Richards, from whence he returned to McGill in 1919 armed with a PH D in physical chemistry. Maass enjoyed his stay at Harvard and continued to correspond with Richards; Harvard was equally pleased with him and he was offered a post there. It was early days in this new symbiosis between physical laws and chemical phenomena and McGill was determined not to lose him. 'Mr. Maass is the only teacher in McGill who is able to carry on instructions and research work in physical chemistry. It would be impossible to fill his place, unless a chemist could be obtained from the United States, which is almost impossible at the present time,' R.F. Ruttan, McGill's director of chemistry, wrote.[7] Maass stayed.

Scientifically, Maass retained a lifelong interest in critical phenomena and changes of state, on which he produced over 180 publications and for which he was elected to the Royal Society of London in 1940. But it was for his pedagogical role at McGill that Maass would be remembered, apart from an unexpected, intense excursion into government science and the military which we shall tell of later. As Steacie was to point out, Maass could be said to hold the same position in Canada that was claimed for Wilhelm Ostwald, who founded 'physical chemistry' and to whom physical chemists everywhere can trace their chemical ancestry. Of course, as a student of both Nernst and Richards, Maass himself fell into this category. But the chemical progenies inspired by Maass's keen scientific spirit belonged to a different generation, one which would play an important role in establishing that subject throughout Canada.

In years to come, in his presidential office at the National Research Council Steacie had the photographs of just two men – Otto Maass and C.J. Mackenzie. Yet as a student, Steacie's earliest memory of Otto Maass did not augur great things. 'You may leave the room,' Maass had told the young Steacie, who could not resist reading the morning papers during one lecture.[8] But if the initial encounter was less than promising, Steacie's later feelings were of pure devotion. For his part, Maass came to regard this young but already confident figure, first with the protectiveness of the teacher, then with pleasure as the protégé became a fully fledged scientist – accompanied perhaps by a very human twinge of envy as Steacie's stature and popularity with the students grew – and eventually, with unstinted affection and respect. After their paths diverged in 1939, Maass generously took every opportunity to initiate procedures for honours for his former protégé, including the tedious canvassing for Steacie's election to the Royal Society of London and an honorary degree at McGill in 1952, which he no doubt felt was long overdue.[9]

Student days at McGill were marked with the usual amount of frivolities, but Steacie also worked extraordinarily hard. He found work absorbing and that it compensated for the 'vaguely unhappy' and solitary life at home. But more critically, hard work enabled him to win all the scholarships necessary to see him through university. It was around this time that Steacie came 'under the spell and stimulation' of Maass who taught the physical chemistry course.[10] In 1921 Steacie was awarded a War Memorial scholarship by RMC and completed his bachelor degree in 1923 with first-class honours and the British Association Medal.

This four-year engineering course (which Steacie completed in three) required, in the first year, eight hours a week in mathematics, drawing and lettering, physical education, and English. In addition, there were physics and mechanics, rounded off with lab work and shop methods. By the second year, calculus and general chemistry were added as well as courses on mapping, construction materials, and surveying. 'The duties of a chemical engineer,' the McGill curriculum stipulated, 'require him to be conversant with chemical processes and the installation of chemical units, and to understand the construction of buildings, the installation and operation of machinery, etc.' In the third year time was divided about equally between chemistry and engineering and during the long vacation students were expected to work for at least six weeks in some chemical industry or equivalent laboratory. By now Steacie knew that he wanted to pursue chemistry – physical chemistry – rather than engineering. His master's degree in chemistry came a year later. The problem of financial support for this and his doctoral research was met, first by departmental demonstratorships for two years, then by a National Research Council studentship, awarded in his final year, 1925–26.

Steacie's doctoral work (on the solubility of gases in metals) was not, however, directed by Otto Maass but by Professor Frederick M.G. Johnson, 'a rather interesting, handsome, urbane man' who in addition to his other accomplishments was a cartoonist.[11] Johnson had graduated in science from McGill and, in the familiar pattern of the day, proceeded to his chemistry PH D in Europe, which he obtained in 1908. By the time Steacie entered the department, Johnson was a full professor; he became head of the department in 1929. Despite these achievements, he was apparently happiest when, 'free to wander afield in company of some artist friend,' he could paint the landscape and enjoy the emancipation of a semi-bohemian existence in the countryside.[12] These interests were manifestly not shared by his student Steacie, who remarked on the strange predicament of finding himself working under Johnson as the result of a 'misunderstanding.' Clearly, Steacie had less time for this artistically gifted chemist than for the more academically inclined Maass; he felt little intellectual loss when Johnson left the university at the early age of fifty-four to lead a life of wealthy leisure. Even the topic of Steacie's doctoral research failed to leave any lasting effect on him, and his interests quickly turned elsewhere. Johnson's influence on Steacie thus appears to have been minimal and quickly forgotten.

Steacie's opinion of his thesis director remained persistently small. Many years later, writing to Otto Maass from Ottawa, Steacie expressed alarm at a piece of news he had just learned from Fred Banting: 'He [Banting] told me this morning that McNaughton was more or less sold on the idea of getting F.M.G. Johnson as a sort of one man committee to do the job. Whether this is only as a sort of personnel supervisor, ... I don't know. Personally I think he would be about the worst possible man for the job, but perhaps I am prejudiced.'[13] And yet, according to a colleague in the McGill faculty of medicine, Fred Johnson had possessed 'so orderly a mind and so critical a judgement' accompanied by 'incomparable excellence' in scientific attainments – a product of a good fundamental training 'combined with clarity of vision and high quality of thinking.'[14] Great attributes, yet none apparently touched Steacie in any way. At the age of twenty-five Steacie was showing exceptional independence, intellectually and spiritually.

In the mean time, the fortunes of the McGill graduate school in chemistry were progressing, with Otto Maass leading the fold. Initial growth, however, had been slow. By 1920 the number of doctoral degrees granted had jumped – but only to three. The year Steacie obtained his doctorate, 1926, established a new record for the university with seven PH DS. In fact, Steacie found himself initiated into the new order disguised as a dentist, since the university did not have sufficient PH D hoods for all seven candidates.[15]

As was customary then, and still is, fresh PH DS anxious to pursue academic life served their apprenticeship by doing post-doctoral research, in Steacie's case with the help of a Sterry Hunt fellowship. Two years later, in 1928, Steacie was appointed lecturer in chemistry at McGill and began his life as a professional chemist. Although his student days were now well and truly behind him, memories of the years of financial struggle to meet the demands of research with the paucity of funds available to science left a deep impression. When he later found himself in the position of leader, these first-hand experiences would influence his policy directions.

As a teacher, Steacie was in his element. After an initial and normal bout of apprehension, due largely to a streak of shyness and social reserve which he always retained, he found the students affable and willing. Steacie knew his subject well and was invariably well prepared; only rarely did he need to refer to his notes. His lectures proved excellent – clear, logically developed, and to the point. They became a topic of conversation among the student corps, who found Steacie's perpetual habit of introducing a sentence with 'actu...ally' or 'as a matter of fact' quite irresistible. Class attention was sometimes heightened by a pool of counts as to how many of these expressions could be found in any one period. The record appears to have been twenty-three 'actually's' and twenty-five 'as matter of fact's,' numbers which dropped dramatically after Steacie caught his students mimicking the expressions in the laboratory. Steacie's class began at the precise appointed hour, developed systematically, and as the clock struck the final moments, was completed from the doorway as Steacie disappeared from view, no doubt anxious to return to the sanctuary of his laboratory.

This was situated not in the chemistry block, where space had been depleted, but in the basement of the biology building which once housed the medical faculty. Steacie shared space – or, as one former student put it, 'a cockroach infested hole' – with a fellow chemist Frank J. Toole and some additional occupants. Actually, it was only a semi-basement and the windows looked out onto the sewer pipes outside, along which, in idle moments, Steacie and Toole could watch the rats hurrying about their business.

Inside, Steacie and Toole went about theirs. Zoology was not really one of their main concerns, but Steacie, in true scientific spirit, made a few empirical observations. To a new student arriving at the lab, Steacie would display a dilapidated Oxford University Press text and explain solemnly that these rats particularly liked the glue; for contrast he would pull out a badly speckled light-blue Cambridge University Press publication, explaining that the cockroaches were very partial to the dye and that he was preparing a scholarly paper on this curious subject. To confuse the student further, there was a conducted tour of the laboratory itself. The larger room, which boasted a cement floor, was 'where they used to lay out the stiffs,' while the smaller labs, with a wood floor, was where the cutting up took place. This was by way of rationalization should any dismembered spirits appear at night. If a student was not thoroughly put off by these revelations, he was sure to get on well. And he usually did, remembering little about spirits that might have roamed but a great deal of Steacie's 'puckish humour.'[16]

Stories abounded then and later about this bright young professor and his talents for doing away with, to his mind, unnecessary regulations and ridiculous pronouncements. There was the tale of how Steacie saved another busy chemist the job of being his project assessor. This was purported to have occurred after he had won a Royal Society of Canada (RSC) fellowship. Under the rules, he was to elect two RSC members as directors of his work and to consult one of them before making changes. Accordingly, since he was by then a fellow of the society, he elected himself as his chief director and after two years presented the necessary document and report addressed to and from Steacie, remembering to thank himself for the excellent report. Then there was the story of how university regulations required staff members to itemize and evaluate all equipment and apparatus annually, with an allowance for depreciation. After a number of years, Steacie was delighted to find that many a piece of lab equipment (in those days as likely as not to have been home-built) could be written off. When it came to leaving McGill this proved a great boon, since he was able to take his apparatus with him and thus carry on the research with the least amount of disturbance.

Most of the time, however, was devoted to serious and legitimate research. Steacie and Toole found time to co-author a paper on the growth of single crystals of silver.[17] By now Steacie was the co-author of a half a dozen publications, including a monograph on laboratory exercises, some joint papers with Professor Johnson, and one with Otto Maass covering diverse aspects of physical chemistry. But it was not until 1930 that Steacie's real chemical interests began to focus, with an investigation into the kinetics of methyl formate decomposition. As we shall see, the mechanisms of how chemical reactions actually occur became an abiding interest which would last all his life.

Steacie and Toole, who was a few years older, got on well together; when not absorbed in their respective researches they would find themselves arguing on equally intractable matters such as life, religion, society, and politics. Toole was then a practising Catholic. Steacie, in contrast, had found himself unable to sustain the kind of religious observance he had experienced as a child. Then, he had had to go to church every Sunday 'no matter what,' as his Anglican mother decreed. He had puzzled over the apparent contradictions between reality and professed religion. If people really believed what the church taught, how could their behaviour be so manifestly converse? Surely, if their religion was really as important as they maintained, they would practise what was preached?

These familiar arguments occupied the thoughts of a young and as yet untried mind. Perhaps maturity, together with an increased awareness of life's failures as well as successes and the limits of science, would have prompted a deeper and less simplistic interpretation. The years ahead, however, furnished little time for dwelling upon such recondite matters; besides, nothing in adult life appeared to contradict his earlier conclusions. Of Steacie's immediate peers who might have influenced him, none seemed to have held deep religious convictions. Somewhere along the journey, Steacie gave up believing in a diety and making Sunday pilgrimages. Growing up in Montreal, however, would leave its mark. Steacie retained an abiding interest in the Catholic Church and how it operated.

His view of life was now inextricably tied up with the beliefs of science in general, with the speed of chemical changes in particular and the factor of time in these changes. 'Humanity,' he once told an audience, 'is merely a time-lag in the process of decay!' This time-lag is permitted only because of the loss of chemical equilibrium; were it not so, we would turn into water and a few gases and 'that would be the end of us.' We are protected against this catastrophe and manage to stagger along for seventy years or so merely because we decompose rather slowly. 'Life on earth,' he reflected, 'is a transitory thing and will not last indefinitely.' Temperature changes over aeons of time would see to that.[18]

The debate over scientific and spiritual beliefs, like the proverbial chicken and egg, has had no beginning and no end; the making and breaking of links between them have formed the very fabric of historical, philosophical, and social matters in every culture. In the twentieth century, the dialogue between divine laws and the laws of nature had reached a new impasse. Still, even to the least cogitative of minds, the fundamental questions, the similarities and disparities, are undeniable. Up until this time in his life, Steacie's world, the society which had moulded him, was the ardently Protestant, Anglo-Saxon setting of Montreal's Westmount, imbued with all the virtues, prejudices, and values of the old world out of which it had grown. As an adolescent, Steacie had not interacted much with other cultures in his society. But now he was able to perceive his heritage with unimpaired vision, not necessarily complacently, but with a pragmatism that would endure for the rest of his life.

The Westmount milieu carried over without interruption to his alma mater. To Steacie, McGill would always stand 'as the highly visible symbol of the English language community in Montreal, the epitome of its history, achievements and wealth.' From the thirties onwards there would be changes, in keeping with the changing times throughout the land, and in the fifties these academic changes would actually accelerate. Outwardly, even in 1962, when F. Cyril James resigned as principal after some twenty-three years, the McGill board of governors still comprised the men drawn from the financial establishment of Montreal: the railway, the banks, the manufacturing and mining industries. The old rule which decreed that all members should be Protestant laymen had disappeared from the statutes, but, as Frost points out, in reality the tradition was maintained. 'There was the occasional lawyer, or physician, but there was no woman, no French-Canadian, no Roman Catholic, no Jew.'[19] As for the social issues which would become a major part of both national and international life, the age of the 'not so quiet revolution' was not quite yet, not at all in most cases.

Steacie would remain a free spirit all his life. Long before it was popular to do so, he would explain to an American chemist why he thought French-speaking Canadian scientists should have the choice of publishing papers in French. And on the occasion of a congregation at which speakers warned students of the tough times ahead, Steacie was observed to be glowering in stoic silence. When it came to warning the young generation of the wisdom of heeding niceties and regulations, it was all too much and Steacie burst out, 'I defy anyone to name a person who has achieved anything worthwhile by toeing the line or following convention.' Nevertheless Steacie, like everyone else, was bound to a large degree by the traditions of the times. The changes and innovations he is remembered for today were achieved not only through his natural abilities, but also within the context of the times in which he lived.

An activity from his past which occupied some of Steacie's leisure hours during these years was the militia. After leaving the Royal Military College he had maintained contacts and was commissioned into his father's former unit, the Royal Montreal Regiment, of which he remained an officer for many years. This took up one evening a week and was a pleasant enough commitment. It came to an unexpected end over an incident then much in the news – the abdication of Edward VIII. Steacie rather took the view that the king was being unfairly treated over the issue of marriage to the divorced Mrs Simpson, and made a stand by writing to his senior officer, who for his part, considered it disloyal to question the constitution and reprimanded Steacie accordingly. It was time, Steacie decided, for a parting of the ways. Also by now time spent with the regiment was becoming as increasing drain on his time.

By far the greatest change in these years of Steacie's life had come with his marriage in 1928 – immediately after her divorce from a former marriage – to Dorothy Catalina Day, the source of his domestic happiness for the remainder of his life. Few can doubt that, without her, his career could not have succeeded so completely. For Steacie, the bright and high-spirited Dot was to replace the experience of his solitary childhood and become the centre of the warm, orderly household which welcomed him home at the end of a long day. In later life, it was one of Steacie's personal pleasures to invite friends and soon-to-be friends home to dinner. Of the large numbers of figures who passed through the portals of the National Research Council, many came to visit and some became close personal friends.

The young Steacies settled happily into married life in Montreal. Steacie's beginning salary at McGill was about $2,500, which would increase to $3,000 in 1930 despite the Depression, when he was made assistant professor. 'Those of us who are familiar with Dr. Steacie's work place a very high value on his services,' F.M.G. Johnson wrote in recommending Steacie's twofold promotion.[20] Even so, it was not a great fortune for one with family commitments; at the end of the month things could become a little anxious. Then, Steacie would find himself taking a brisk forty-minute walk to the university rather than taking the bus. Of course, there were also ways and means to combat these deprived years. Steacie, by now an habitual smoker, would remember to put all his cigarette stubs in a container. When the cigarettes ran out, he would revert to pipe smoking by removing the paper and reclaiming the tobacco.

By now, the family was complete with a daughter, Diana Jeannette (Jan), and a son, John Richard Brian (Dick). The four made their home in various rented houses, some of which proved very grand indeed, but for the fact that economy did not allow the whole house to be heated. Generally, family life ran smoothly with the occasional peak and trough. On one such occasion, which Dot now recalls with amusement, when the family had to make a move Ned undiplomatically developed mumps, much to everyone's annoyance. Still, it was his first bout of illness since their marriage and had to be forgiven. The Depression years lingered on. Even the children were becoming accustomed to the fate of the many unfortunate. Young Dick amused his parents greatly by inquiring in kindly tone of a retired neighbour tending his garden, 'And are you still out of a job?'

Scientifically, these were very productive years for Steacie, with a prodigious output of thirty-six publications between 1930 and 1934, mostly co-authored. His interests were still broad and ranged over various physical chemical properties, such as solubility and adsorption as well as the problem of reaction kinetics. Steacie's personal enthusiasm for research was now well established and students gravitated toward his basement laboratory; within a few years Steacie found himself directing an increasing number of graduate students. Between 1930 and nine years later when he left McGill, at least twenty-four masters and doctoral candidates had come under his guidance.

These were happy years and intellectually fulfilling. He was doing what he loved most, and though departmental paper work was unavoidable, it was kept to a manageable level. There were the usual weekly departmental meetings, for example, at which he acted as secretary and took minutes. Steacie shared the teaching of physical chemistry with Maass, the latter being responsible for the more advanced course and Steacie taking thermodynamics. These lectures took up four sessions a week together with an additional nine hours of physical and colloid chemistry laboratory work.[21] Textbooks chosen for the advanced course were the familiar Nernst's *Theoretical Chemistry* and H.S. Taylor's *Treatise on Chemistry*. Maass and Steacie both felt that a more elementary introduction to physical chemistry was needed and collaborated to produce a text which appeared in 1931. In 1939 a second enlarged edition appeared with new chapters on atomic structure, thermodynamics, colloidal properties, and a completely revised chapter on rates of reaction.[22] Nothing shows more clearly the differences brought about by these eight years. The discoveries of the previous decade opened up many new areas and offered glimpses of new horizons. But before these new horizons could be approached, much was to ensue for Steacie personally.

2
A Taste of Europe and the Years After

In the spring of 1934 a cherished dream that would focus his ambition for the rest of his life came true for Steacie. News came that he had been awarded a Royal Society of Canada fellowship, to spend a year in Europe to work with K.F. Bonhoeffer in Frankfurt, Germany, and A.J. Allmand in London, England.[1] The sum was not ungenerous, about $1,500, close to half his annual salary; still, with a family and travelling expenses to consider, it would be tight. With an additional sum borrowed, the family set sail in June aboard the Cunard ss *Ausonia*. The weather was fine, as was the experience, although Dot and the children succumbed to seasickness. The family took the train from Cherbourg to Paris, marvelling along the way at the novelty of seeing the continent for the first time, relishing the pleasures and pains of all travellers faced with strange experiences. To the Steacies' unaccustomed eyes, Europe in the thirties was a world apart; people's attire seemed so different from those back home and the houses and narrow streets looked much as they did in paintings. By the time they reached Paris everyone was exhausted and indulged in a long rest before venturing to explore the Champs-Elysées and the Eiffel Tower. Their destination, Frankfurt, was not too far from Paris, a reasonable train ride away.

Just a few days later the four travellers found themselves at night at the railway station in Frankfurt. A friendly taxi driver recommended that the family not bother tramping the city in search of accommodation but just go across the street to the nice little hotel in the town square. Sure enough, the Steacies found a warm welcome awaiting them at the Hamburger Hof. There and then, they decided to stay put for the remainder of their time in Frankfurt.

Frankfurt was an attractive city which had retained its dignified central

A Taste of Europe and the Years After

area by spreading south towards the river; there was also the beautiful Altstadt with its famous Weinstuben which students from the university frequented when funds permitted. In the thirties, Frankfurt was cosmopolitan, a highly cultivated city with great industries, and home to a wealthy, largely Jewish, upper class.

Not for nothing were the centres of excellent science to be found in Germany in the twenties and thirties. The chemical laboratories at the University of Frankfurt were magnificently equipped. Analytical procedures were already sophisticated and glass apparatus – the essential equipment of every chemist then – already had such luxuries as standard interchangeable ground-glass joints and Jena sintered glass filters. Young enthusiastic chemists from everywhere, but especially England, found these 'quite an eye opener,' as chemist Alexander Todd put it. By 1929 organic chemists were treated to microanalysis as a routine service and the business of catalytic hydrogenation with platinum or palladium catalysts at standard temperatures and pressures was quite the norm. Not until many years later did similar analytical wares find their way to laboratories abroad, carried there by returning scientists. Across the road from the organic, inorganic, and pharmaceutical laboratories stood the physical chemistry domain, presided over by Karl Friedrich Bonhoeffer.

Born in 1899 in Breslau, young Karl soon moved with his family to Berlin, where his father became professor of psychiatry at the university. As a student, Bonhoeffer came under the influence of Walter Nernst, Max Planck, Albert Einstein, and Fritz Haber. Like all who passed that way, Bonhoeffer would have inherited from Nernst the ability to design simple but elegant apparatus, adapted for specific analysis, which required new physical insight into old chemical problems; and from the eminently practical Haber, the director of the Kaiser Wilhelm Institute for physical chemistry, he gained a knowledge of chemical technology. Armed with these abilities, Bonhoeffer, by applying quantum mechanics, would discover with Harteck (in 1929) an allotropic form of hydrogen. At the Haber Institute, where he worked from 1923 to 1930, the young chemist produced a string of excitingly original work, including an extensive series of investigations into the chemical properties of atomic hydrogen. As we shall see, atomic reactions were the reason and purpose of Steacie's pilgrimage to Germany.

In 1930, Bonhoeffer was pleased to take up the chair of physical chemistry at the University of Frankfurt, though he did not particularly relish the director's duties, which included a heavy load of administration, teaching, and the awful ordeal of student examinations. But it would be a constructive time; Bonhoeffer began to study chemical reactions using labelled hydrogen. The potential of these techniques would lead him eventually to physiological chemistry and biological physics which increasingly absorbed his interest.[2] It was in the midst of these developments that Steacie joined Bonhoeffer at Frankfurt. By now Steacie was familiar with Bonhoeffer's publications in the *Zeitschrift für physikalische Chemie*, the *Zeitschrift für Elektrochemie*, and other leading journals of the day, which undoubtedly speeded the program ahead.

As a more senior visitor to Frankfurt, Steacie was saved from much of the tortuous excesses of German academic life, such as the trial of an oral examination to which all doctoral candidates were subjected. Such an experience has been known to cause damage of a permanent nature, leaving the candidate, even if successful, under the illusion that tradition required him to treat all subsequent students and subordinates in like manner. Of the many North American students who trained in Germany in late nineteenth and early twentieth centuries, it was expected that a few would carry back with them, and worse, perpetuate this overblown role of the Herr Professor. Of course, there were always exceptions, of which Bonhoeffer was one, a fact Steacie appreciated. However, even as a young man Bonhoeffer had acquired the familiar traits of the absent-minded professor. At Frankfurt, to counteract the impossibility of knowing everything on subsidiary subjects required for the 'doktor examen,' the candidates developed a curious but usually effective ritual which Alexander Todd describes in his reminiscences. 'About a week before the date of the examination one called on each of the examining professors at his home bearing a bunch of flowers for his wife and had a chat.' The object of this exercise was to acquaint the said professor with the relevant ('interesting') areas on which the candidate wished to be examined, and thus avoid embarrassment to both parties during the examination. In Todd's case, it did not work out too well. Bonhoeffer, who had only just arrived in Frankfurt and was still unaccustomed to examining, confused what Todd had described to him a week before as interesting and uninteresting. Accordingly, Todd suffered 'a rough half-hour on predissociation spectra and some other photochemical topics' from a lecture course through which he had usually slept.[3] The fact that these were the very topics which would bring Steacie thousands of miles to the same destination merely shows that one man's scientific passion was another's downfall, although Todd apparently developed no permanent scars.

When Steacie arrived in Frankfurt, the Physikalische Chemisches Institute, located on Robert Meyer Strasse, had been completed only two years earlier and, as Steacie noted with satisfaction, 'is just about perfection in the way of all kinds of equipment.' He was full of enthusiasm for the novel experience of a European university and noted the privileges of this particular one. 'Bonhoeffer is a very nice chap and the laboratory is run on entirely non-German lines. Everyone does what he wants when he wants without too much of the "Herr Direktor" business,' he wrote approvingly to Otto Maass. The residual German custom of formality was, nevertheless, quaint for one used to the less formal ways of home. Correct procedure remained an important part of the day's business. Each morning the occupants of the laboratory (three others apart from Steacie, including K.H. Geib) would greet each other solemnly with 'Morgen,' a click of heels and shaking of hands; after lunch, more greetings, click of heels and hands shaken again all around.

It was a happy and productive group. 'I am sure that it is going to be a very pleasant year, and that I will get what I wanted out of it,' Steacie wrote.[4] He began his investigation by carrying out reaction rate measurements in heavy water, a curious beginning which would have interesting repercussions in the perilous days of war ahead. 'They [the Germans] are going in for heavy water very strongly and are making it in large quantities,' Steacie observed. Things went well; the work on atomic reactions turned out to be extremely interesting, 'just the sort of thing I came here for.'[5] By the beginning of December the reactions involving deuterium and water were completed and work turned to the reaction between deuterium and ammonia.

Meanwhile Dot and the children were also finding Frankfurt to their liking. In the morning, Dot taught the children their lessons and in the afternoon the three would explore the city, finishing off in satisfaction with a cup of cocoa and a German kuchen. The children, now aged seven and nine, had become great friends with their landlady, Frau Hellwig, and were allowed the privilege of watching activities in the kitchen. Sometimes, very early in the morning, they would accompany the staff to the local market for fresh produce, perched on bicycle handlebars. Jan and Dick were soon chatting away happily in German, confidently acting as interpreters between new guests and the staff, and generally taking command of things. In the evening, while Frau Hellwig, now a firm friend, kept an eye on the children, Dot and Ned strolled out and savoured the city's charm. One particular incident stood out which, years later, took on added significance. Two uniformed ss officers accosted Dot and a young acquaintance on one evening outing, to no avail; it was enough though for a hearty laugh all round on returning to their escorts.

The Steacies were entertained to dinner by the beautiful and accomplished Frau Bonhoeffer, Greta von Dohnanyi, the daughter of the composer and conductor. Greta was herself very musical and Dot greatly appreciated her recitals although Ned was not much attuned to music. On first encounter, Dot was surprised to see their hostess smoking a cigar. It was one manifestation of the confident, sophisticated generation which had grown up in the decade after the First World War, when Germany had experienced a defiant revival. The passing of a decade had worked wonders on those with short memories. By hard work and enterprise, German industry had recovered, expanded, and was proving a formidable force in Europe by the mid-twenties. People were rapidly regaining their confidence after the misery of war and the devastations of defeat. By the self-confession of one Berlin resident in that era, 'the young men and women who had grown up in the First World War and in the post war years were an active and enterprising generation, free of all illusions, but also without any ideals. The individual effort was entirely devoted to personal interests in science, art, literature and similar pursuits ... We had created a world of intellectual achievement, accompanied by a sound cynicism on all emotional issues.'[6] While the lack of ideals did not extend to all, intellectual achievements there were indeed. This was the age of Max Planck, Einstein, Pauli, Schrödinger, Heisenberg, and Max Born, to mention but a few; the world would wait a long time before such a plethora of creative thinking could burst forth again. In the opening decades of this century Germany stood, as Steacie observed for himself, at the pinnacle of scientific achievement; it was all the more tragic that politics would soon thrust everything before them into an abyss.

That summer of 1934 the family spent a month touring the country, visited the Island of Sylt, and played tourist in Hamburg. Over the Christmas holidays, they visited Munich, Garmisch, and Nuremburg, becoming 'fairly accomplished third-class travellers' in the process. Their travels convinced Steacie that Germany was 'quite a pleasant place to be in these days,' although there was some anxiety over their finances after it was announced that Germany would abandon the Registered Mark.[7] In reality, of course, all was not well. Hitler had already been in power over a year and although things would not get out of hand until a few years later, rumblings could be heard by those who were listening. Meanwhile, Bonhoeffer had been offered the chair of physical chemistry at Leipzig, a singular honour to follow in the footsteps of the great Wilhelm Ostwald. Under Ostwald's confident and pioneering guidance, Leipzig in the later part of the last century had become the outstanding centre of the new physical chemical studies, and its prestige had been maintained by the presence of both Arrhenius and Nernst. Nevertheless, Bonhoeffer hesitated. The offer had come some time in late summer and in September Steacie told Otto Maass that 'Bonhoeffer has practically definitely decided to turn down the Leipzig job.' But by mid-December the situation had changed and everyone prepared to make the move to Leipzig. It was a cause for regret since the present laboratory was so marvellously fitted up and equipped. 'In fact,' Steacie observed, 'apart from unnecessary ornaments, it would be impossible to find a better laboratory anywhere.' Part of the problem would be the dismantling of the carefully constructed apparatus, much of glass. Still, Steacie regarded the forthcoming upheaval with his usual cheerful pragmatism. 'I will be rather glad to go there on account of the much greater importance of the university. I am afraid, however, that the institute there will not be quite up to Frankfurt.'[8]

For Bonhoeffer the move was to prove a fatal mistake. Leipzig would soon be caught up in a chapter of history which would leave the entire Bonhoeffer family devastated. After Hitler became chancellor of Germany in 1933, Karl's brother Dietrich, a theologian and passionate defender of his beliefs, would find himself taking on the Führer and the pastors of the German churches, who were also state officials, all at once. Dietrich was not the only member of the Bonhoeffer family to be 'lost to the Third Reich.' Karl was spared death but suffered harsh Nazi persecution, an experience which later he would not discuss. Only in the post-war years would Steacie learn of the tragedies which had befallen his much respected mentor. But even in 1934, Steacie sensed that things were happening within Germany. 'The situation in German universities today is rather strange,' he remarked; 'they have a tremendous amount of equipment, technicians, assistants, etc., but are becoming very worried over the lack of students. The attendance of all the German universities has been falling steadily' and, he added innocently, 'of course, it is very nice for the men as there is very little trouble in getting jobs when they get through.'[9] Clearly, Steacie's observations were limited to the small and harmonious group which worked under the gentle eye of Bonhoeffer.

By now, Steacie's German was sufficiently proficient to participate in scientific discussions and he felt confident enough to give a colloquium. But like any visitor to a new country, the nuances of daily life for those caught up in this complex situation could not be assimilated in a month, or even a year. Travelling around the country, Steacie saw great industries, beautiful cities, and apparent affluence, mingled with outstanding scientific and intellectual achievements. Steacie attended lectures by Planck and other notables. More significantly, all this was seen in the light of the destructive depression which still lay over his own homeland, memories of which were only too real. During their stay, the family observed scenes which signalled things to come, such as storm-troopers parading through the streets (carrying spades rather than guns). They listened to Goebbels a number of times and attended one of Hitler's famous rallies – an alarming experience – more for the sight of the crowds than the impact of the Führer's impassioned words. The whole family found Hitler very noisy.[10] Steacie's impression of German politics from these brief encounters cannot be said to be very astute, which is perhaps not surprising, culled as they were in the main from within the laboratory. And they would have mattered little but for a surprising event which took place after the family's return to Montreal.

In the mean time, after the friendly environment of Frau Hellwig's establishment in Frankfurt, the family was not finding Leipzig a very congenial experience. So they were all pleased when Steacie discovered that his work could be finished by April. The four travellers set off in eager anticipation on the next stage of their journey, to England, embarking upon a little more exploration of Europe along the way.

The summer of 1935 was unusually dry and hot throughout Europe – a good year for wine and tourists, as someone remarked. The Steacies arrived in England to find the countryside almost scorched brown. It was Jubilee year for King George V and Queen Mary and London was decked out in festive mood. For Steacie it was to be the first of many visits; there would be plenty of other occasions to discover that dry sunny days were not always the norm in England. Fortunately they had beaten the tourists by a slim margin and managed to acquire a small but pleasant apartment in the heart of Bayswater, complete with sitting-room and kitchen. Dot and the children entertained themselves watching the parades and the processions in and out of Buckingham Palace, with the small princesses waving regally from their coaches, the fireworks, and much waving of flags.

Steacie took the London underground (which was very convenient once you understood the intricacies of the various routes that darted off in all directions) from Bayswater to the Strand, where King's College was located. Being in the hub of the city had its blessings and its drawbacks. The hustle and bustle of London were all around; a short walk brought the explorer to Trafalgar Square, famous for its pigeons, fountains, art galleries, and St Martin-in-the-Fields. The disadvantage was the traffic and, if scientific experiments involved sensitive measurements, combatting the constant vibrations caused by the underground which ran close by.

Steacie's destination, the physical chemistry department, was headed by Arthur J. Allmand (1885–1951) who had trained at the University of Liverpool under F.G. Donnan. Fresh with enthusiasm for the new discipline of physical chemistry as he had learnt it from the grand masters van't Hoff and Ostwald in Germany, Donnan was gifted in the art of proselytizing. Under his tutelage, the little group at Liverpool, including Allmand, 'became ardent physical chemists on the spot.'[11] Donnan's group thus received a broad European outlook without necessarily having left their native shores, but which would nevertheless influence their thinking for the rest of their scientific careers. No doubt at the instigation of Donnan, however, Allmand did make a pilgrimage to Germany, to Karlsruhe, for a year with Haber, and subsequently to Dresden. The young Allmand began his chemical career with electrochemistry, following in the wake of the exciting new theories of Debye and Hückel on dissociation. But by 1925 his interest had focused on photochemistry and the peculiarly elusive reaction between hydrogen and chlorine. This deceptively simple combination had puzzled chemists for an inordinately long period of time and was to become one of the most studied of all chemical reactions. By this time Allmand had joined the University of London's King's College as professor of physical chemistry, a post which he held from 1919 until his retirement in 1950. Before arriving in London, Steacie would have been conversant with Allmand's growing reputation in the field of reaction kinetics. As research director, Allmand acquired a habit, aimed at developing students' reasoning powers, of offering few opinions but proffering probing questions, such as 'What do you think is the mechanism'? Students, invariably young men full of enthusiasm, were thus induced to propound their own favourite theories, 'which often were obviously absurd,' if only they could have realized it. The subject of reaction mechanism – the purpose of Steacie's sojourn to London – was an intractable chemical discipline which had burst upon the chemist's horizon in the second half of the nineteenth century (although not then recognized by that term) and quickly established itself as stubbornly insolvable in as many cases as not.

In Allmand, as in Bonhoeffer, Steacie found a mentor with much to admire. The serious, bespectacled professor was a 'glutton for work,' toiling into the early hours of the morning. By all accounts he was an excellent teacher whose lectures 'were a model of clarity and the pains he took were beyond belief.'[12] Steacie's stay at King's was brief, and probably centred around discussions on new techniques and information coming out of the rapidly expanding chemical literature. If there had been time for personal discussions Steacie would have discovered (surprisingly) that the somewhat physically delicate Allmand had great enthusiasm for army life; that (less surprisingly) Allmand, converted at the time of his marriage in 1920, was a devout Catholic and in all that he touched the abiding conviction was a duty to 'his country, his university and his students to the greater glory of God.'[13] As director of the laboratory Allmand was quiet, conscientious, tolerant, and efficient; Steacie regarded him highly. In 1949, when Steacie had outstripped most photochemists of his generation with a complete mastery of the subject, he would return to King's to lecture on gas phase reactions of organic radicals, the very subject Allmand now excelled in, at the invitation of the professor. Allmand, it seems, had a favourite dictum which, as enunciated in his memorial, reads 'a professor of one of the natural sciences ... might be a great researcher, a great administrator or a great teacher. Sometimes he might shine in two of these categories but to do so in all three [would be] unattainable for most people.'[14] That Steacie managed to encompass all these, if not at the same time, would have pleased this grave and unassuming chemist. Another pleasing aspect of the sojourn at King's was the acquaintance of Harry (later Sir Harry) Melville, with whom Steacie shared the laboratory. This would develop into a friendship which deepened over the years as both men found themselves in the role of administrator.

That summer of 1935 an incident much occupying the country's scientific community caught Steacie's attention: the detention by the Russians of physicist Peter Kapitza. This was not the kind of detention meted out to dissidents as we understand it today, but in a way signified the beginning of a new world order. Of the many jewels which sparkled in the brilliant scientific crown of Rutherford in the Cambridge of the twenties and early thirties, few shone more brilliantly than Kapitza. This young Russian physicist had arrived at the Cavendish in 1921 and was to become the most important figure in Rutherford's life. During these years Kapitza habitually spent the long vacations in Moscow, visiting his mother and relatives. In the summer of 1934 the Russians refused to allow Kapitza to return to Cambridge, despite repeated efforts by Rutherford and other internationally renowned scientists to reverse the decision. It was an unhappy affair for the aging Rutherford. Nevertheless, in his letter to *The Times*, published in May 1935, he stoutly maintained that 'science is international and every scientist hopes it may remain so.' In principle, Rutherford believed that it did not matter where Kapitza did his work, provided he had the right equipment and atmosphere to continue his creative thinking. Rutherford appealed to the Soviet government to permit his former protégé freedom of movement. The president of the Royal Society, Gowland Hopkins, echoed these sentiments and wished Kapitza's services 'to be rendered wherever they can best promote the interests of pure and disinterested science as an international pursuit.'[15] It was a singularly generous statement, coming from the institution which had recently paid out munificent sums to build the Mond Laboratory in Cambridge specifically for Kapitza. In none of these incidents making waves on the normally calm backwaters of academia was Steacie involved in any capacity, except to observe the convictions of these men of science. But just over twenty years later, the same events would take on greater significance. As president of the NRC, Steacie would receive a rare honour and be elected to membership of the Academy of Sciences of the USSR, would be received by Kapitza in his own home, and would observe for himself the Soviet mode of science. Of greater significance still, Steacie would have cause to ponder the meaning of international science in the midst of the cold war, and whether and how this internationalism could affect the outcome of world politics.

The months in England were flying by. Steacie's work satisfactorily concluded, friends persuaded the family that their last month should be spent in western Scotland, in the secluded Kyle of Lochalsh. The experience, however, proved not quite what they had anticipated. The train journey to St Andrews in Inverness was interesting enough and the children looked hopefully for the Loch Ness Monster which, alas, 'was not there.' Arriving in the depths of the Scottish countryside, the four accomplished travellers found awaiting them mountains and sheep, and more mountains and sheep, and rain and more rain. After the first day or so, even the mountains disappeared in a shroud of mist. It rained as only it can rain in Scotland, every day for the whole month of their holidays. Steacie tried to get up a game of golf with a new-found friend. The rain, however, had been excessive, even for the natives; a smart swing of the club and the ball disappeared with a splash into the bog and was not seen again. Defeated, the two indulged in a 'wee dram' to ward off the cold, the Scotsman having first ascertained that his partner was not a teetotaller. No great harm ensued from this wet interlude; the children became great walkers and they paid a visit to Skye, which was similarly swimming in rain and mist. The whole family learnt the art of drying off in front of a small fire, and enjoyed the local fish, and the two desserts which followed.

So ended the family's first taste of Europe. It had been altogether interesting, as novel experiences usually are. But for Steacie it was to prove something more. For much of the last decades of the nineteenth century and the opening decades of the twentieth, the centres of scientific excellence were in Europe, and in Germany in particular, to which aspiring American (and Canadian) science students journeyed to receive their initiation into modern science. Of seventeen advanced students at the Institute of Physical Chemistry at Göttingen in 1896, no less than six had come from the United States.[16] Steacie had followed in this tradition, although by then graduate schools had been established in both the United States and Canada, thus allowing for European post-doctoral experience. What he observed in Germany and England was to set the standard for new scientific laboratories on the other side of the Atlantic. His stay in Frankfurt would have taught him the indispensable advantage of having an expert glass-blower in the lab, for example, and the way new techniques could revolutionize scientific explorations. At the human level, he would have observed the personalities of two of the foremost photochemists of the day at work – and noted the impact their leadership imparted to members of the laboratories. Above all, Steacie would have measured his own abilities with those he met in the schools of science in Europe and found that the future looked bright. It was a time of growth, in knowledge, in experience, and, not least, in confidence. Even more significant for times to come, the large numbers of students returning to the North American continent would result in new schools of thought and expansion of learning in the sciences. Changes, both subtle and innovative, would take hold in American universities and industries alike. Now they said goodbye to Scotland, still drenched in rain, and arrived back in Montreal in time for the new academic year.

After a long absence there was naturally much catching up to do. Steacie's students, who had more or less been transferred to Otto Maass's care during his absence, demanded considerable attention. He resumed a new teaching load and took up where he had left off as secretary, taking minutes at the frequent departmental meetings, and, of course, he dived back into his research with renewed enthusiasm. During his sojourn in Europe the flow of publications had not dried up as they might understandably have done. In 1934 fourteen publications appeared and in 1935 ten papers, now, with few exceptions, concentrating on gas phase kinetics. Of course, much of these came from work of previous years, combined with students' efforts. Publication schedules, though infinitely less burdened than today, took long enough even then. For Steacie, the two decades after his return from Europe would be the most productive of his scientific career. Over 160 papers and four books, the last one covering two volumes, made their appearance. There would be another 68 or so papers before and after this time, but by the early fifties Steacie's reputation as a leader in his chosen field had reached its zenith. To the end of his life, when great honours and power had been heaped upon him, Steacie would continue to uphold his role as a scientist above all others. Even on glittering social occasions at which only wielders of great power were invited to eye each other, Steacie was remembered as introducing himself as a chemist; he simply liked himself best in that role. For chemistry as a whole, these were also years of consolidation after an exciting era of progress and fundamental understanding of some of the most elusive problems in the entire chemical endeavour.

By December, Steacie had settled back into the familiar freeze of the Canadian winter and the equally familiar academic routine. That his entire experience in Germany had made a good impression on him became clear when he addressed the Montreal branch of the Engineering Institute of Canada. Afterwards, the *Montreal Gazette* reported that Steacie had declared that the 'Hitler regime is stable.' Moreover, 'anyone who objects to Hitler should blame President Wilson and David Lloyd George.' Harsh words, all the more astounding when attributed to the (subsequently) thoroughly apolitical Steacie. Germany, Steacie pointed out, after the Treaty of Versailles, had suffered incredible hardship: unemployment became rampant, the economy had disintegrated into chaos, as had law and order, and the nation had lost its self-respect.[17] At the time, these facts were certainly well known. The treaty, drawn up in considerable haste on 28 June 1919 by the then 'Big Four' (France, Britain, Italy, and the United States), forced upon the vanquished nation a liability 'for sums unspecified and without foreseeable end.' John Maynard Keynes had already stated these realities forcefully at the time of the treaty. Germany would not be able to pay the sums demanded in reparation. Indeed, Keynes pointed out, the problem did not depend solely on promises or economic factors. 'The policy for reducing Germany to servitude for a generation, of degrading the lives of millions of human beings, and of depriving a whole nation of happiness should be abhorrent and detestable ... even if it were possible, even if it enriched ourselves, even if it did not sow the decay of the whole civilized life of Europe.'[18] But the press outside Germany, Steacie maintained, persistently misrepresented these facts. The young professor, recently returned from Germany, warming to his subject of misrepresentation, stoutly maintained he preferred the control of Goebbels to that of the advertising manager of a large store. Whether it was fair to equate the propaganda of advertising with the general press is debatable, but Steacie was to retain a lifelong distrust of all forms of advertising. As for the discrimination against the Jews, which he certainly did not condone, he pointed out that that phenomenon was not confined to the Nazis. Many other countries held similar views when it suited them – even Canada, where it merely took a covert form.

Whatever the implications of this last observation, the impression of the Germany he had just experienced would be the most mistaken one Steacie ever held, and it was not long before he was taken to task. Otto Maass's reaction to this item in the morning press was one of disbelief and, for once, he lost no time in courtesies. Steacie emerged from their confrontation no doubt better informed and suitably chastised. For Maass, the entire history of Germany was one of deep personal concern. 'The emotional shock of World War I and the attendant conflicting German and Canadian loyalties; the escape from Germany; the comparatively early death of his father and the anti-German feeling which forced him closer to his widowed mother,' all told a tale. Maass developed 'an intense Churchillian hatred of the Nazis,' which would not have permitted his younger colleague at McGill to misconstrue the new situation.[19] As a boy, Maass had made frequent trips to Bonn with his parents, had close relatives there, and was proud of the Germany he had known in childhood. To see it now destroyed was an experience shared only with those who have suffered a similar loss. In any case, the incident was not mentioned again in subsequent correspondence between Maass and Steacie.

Steacie's visit to Germany coincided with a period when the fate of scientists of whom Hitler disapproved was at its height. This group was not confined to Jews: Bonhoeffer was not Jewish, neither was Nernst, yet both suffered, though to different degrees. Indeed, none of the men heading chemistry branches in the Frankfurt that Steacie knew would escape the Nazi hand. If Bonhoeffer, who most certainly was aware of the events all around him, had enlightened Steacie at that time, the picture would have looked quite different. Signals of things to come had started quite early. In 1933 one of the most notable chemists of that generation, Fritz Haber – the man who had so influenced Bonhoeffer's career and who was another star in that 'brilliant constellation' of science in the

Berlin of the twenties and early thirties[20] – was dismissed from his post as director of the Kaiser Wilhelm Institute at Dahlem. 'For more than forty years,' he told the authories, 'I have selected my collaborators on the basis of their intelligence and their character and not on the basis of their grandmothers, and I am not willing for the rest of my life to change this method which I have found so good.'[21] If any one factor was needed to convince Steacie of the unhealthy state of matters in Germany, this view alone should have sufficed. Few things would hold greater meaning throughout his life than that scientists should be judged by, and only by, the quality of their work. Personalities came a poor second; as for the denomination of the person's beliefs, this would stand for very little.

That summer of 1935, Steacie's contemporary and later long time friend, William Albert Noyes Jr, and Mrs Noyes, also found themselves in Europe, sailing from New York to Antwerp on the German ship ss *Belgenland*. When they were about half-way across the ocean the official German flag was changed from the striped red, yellow, and black to the Nazi swastika. 'My wife,' Noyes recalled, 'felt like jumping off the boat.' For Noyes, who had spent some school years in Germany and undergraduate days in France, there was no doubt that war was inevitable, once Hitler came to power. Steacie's and Noyes's careers were to have much in common. Born only two years apart, while one lost a father in his teens, the other had lost a mother at the age of two. The two years nevertheless made for different experiences; Noyes described himself as a 'Victorian in the 20th century' and had experienced the awfulness of the trenches in the First World War. During those nightmarish days, when there were duty stints of sixteen to twenty hours each day for perhaps three or four days at a stretch, Noyes began the habit of smoking heavily, 'just to keep awake.'[22] It was a lifelong habit which he shared with his friend Steacie, who was spared the trenches by being born just that bit later than Noyes.

It is doubtful whether Steacie and Noyes had yet met, but their respective roles during the coming war would bring frequent meetings. In the United States, Noyes would become chief of chemical warfare and make frequent visits across Canada and especially to Ottawa, where the 'relatively small but extremely competent group of scientists and military personnel' made for relationships which would long outlast the war. Afterwards the race would be on, in the field in which both men aspired to make their mark – photochemistry. Like Steacie, Noyes had become interested in the subject some years earlier and had had occasion on an earlier visit to Germany to discuss problems with Bonhoeffer, then still at the Kaiser Wilhelm Institute at Dahlem. Because of these repeated visits to Germany, and because of early contacts with a wide cross-section of European scientists, Noyes had come to see a different picture from Steacie of the Germany of 1935. Events would soon reconcile their disparate views.

Steacie's apparent approval of the Hitler regime caused little or no ripple of interest in the McGill community, apart from the disapproval of Otto Maass. The campus was deeply involved in debates of its own. The university was surviving the Depression in the best way it could; expenditures were being held to a minimum, salaries had been cut, vacancies left unfilled, students' fees raised. McGill, like most institutions, 'could only batten down the hatches and wait for better times to come.'[23] When Steacie returned to McGill in the fall of 1935, the university was adapting itself to a new principal, Arthur E. Morgan, an experienced administrator recently brought over from England. From the beginning, however, relationships did not prove congenial. Whereas both the chancellor, Edward W. Beatty, and the new principal held the highest ideals, not to mention force of personality, Beatty leaned politically to the right, Morgan to the left. These personal inclinations might have passed unnoticed but for the fact that these polarizations were soon highlighted by incidents on campus. The expression of socialist ideals by a number of professors and students had begun some years earlier. When the chancellor pointed out to Morgan that 'our university socialists are not only preaching socialism constantly, but are more active in trying to induce others to accept their doctrine than any other class of propagandists in the country,' the principal's reply offered little comfort. 'I have no solution for the problem which will I hope always be with us. I hope because as I see it the only condition of its solution is the establishment of an authoritarian state, which God forbid! ... all that we can actually do is to impress on them at all seasons that rights imply responsibilities.' Principal Morgan's downfall, however, came about over a much different, and seemingly unimportant, issue. While in university life 'large mistakes of policy can be understood and in time forgiven,' the impact of apparently small acts of undiplomatic insensitivities cannot. Morgan, for some strange reason, attempted to alter the colour of the McGill letterhead from the traditional red to, paradoxically, conservative blue. Such lack of insight could not be tolerated; besides which, blue was also the colour of that other learned institution, based in Toronto. As Frost puts it, 'even students realized that a man who could paint McGill blue simply had to go.' The matter of the university's 'unfortunate reputation as a centre of socialist propaganda' was soon taken in hand by the man chosen to replace

Morgan, the American Lewis W. Douglas. The remedies prescribed were three: redefinition of tenure; promotion only for 'a list of carefully selected men' which would emphatically exclude the supporters of 'collectivism'; and to counter the more socialist-minded senior staff already tenured contrary voices would be brought to the university by two newly established visiting professorships, one in history and the other in political economy. One such voice recommended was that of the British economist Lionel Robbins. As implemented, the strategies 'proved remarkably enduring' in the life of McGill. More widely, these facets of campus politics in the thirties doubtless reflected the social and political climate of the day which 'even in Canada was super-charged with ideological emotions.'[24]

3

A Chemist Finds His Calling

It is doubtful whether Steacie had much time to devote to politics during the mid-thirties, either on or off campus. By inclination, and of necessity, chemical affairs now dominated his life.

The topic which from now on would become Steacie's scientific passion had promised to be troublesome from the start. Atoms or free radicals wandering around unattached to a proper molecule, even if only for a fleeting flicker of time, outraged all sense of chemical propriety. It was too ludicrous an idea to contemplate and yet had proved too tantalizing to let go. The existence of these curious entities which possessed free or unpaired electrons had been studied and debated since the time of Lavoisier. As Moses Gomberg observed in 1928, 'the existence of radicals ... has been, in turn, surmised, believed in, considered as demonstrated; then the question became debatable; and finally, beginning with 1860, the independent existence of radicals was looked upon as wholly improbable.'[1] It was not until 1900 that Gomberg considered the question of the existence of radicals to have been finally resolved. In that year he successfully isolated the triphenylmethyl radical $(C_6H_5)_3C$.[2] So confident was he that here at last was a chemical treasure to which he held the key that he wished to reserve the entire field of research for himself. This was much too rash. Could Gomberg but have foreseen it, he was far from being the destined expositor of free radical phenomena to a grateful world. The controversies and difficulties would persist.

For a start, the concept of trivalent carbon suggested by Gomberg's radical defied all acceptable rules of the day. Not surprisingly, the chemical community reacted more or less with one accord – total scepticism. Had not the great Wilhelm Ostwald made it clear only four years previously that 'it was finally recognized that the very nature of the organic radicals is inherently such as to preclude the possibility of isolating them'?[3] Besides, even if such an entity could be isolated, how could it be certain that Gomberg's compound was not merely a dimer, a frequent occurrence in organic chemistry? Opposition was loud and strong and increased when molecular weight determination showed that dimers were indeed a possibility. The difficulties confronting chemists did not reside in the reality of free radicals alone, however; these were merely a part of the wider and more intractable problems of chemical kinetics and reaction mechanisms in general.

In the second half of the nineteenth century, another, less dramatic, chemical revolution had taken place in Europe. The realization had dawned that chemical reactions do not, in general, proceed by a simple, straightforward, one-step action. Theory and experiments began to reveal that within one chemical process, any number of accompanying reactions could be taking place, singularly, simultaneously, or in a given sequence. What appeared as the result of a reaction was only the end product of a large number of chemical changes.

The second decade of a new century dawned. Chemists, in search of explanations of how reactions actually occur, found themselves still floundering for lack of a sound theory. The nebulous radiation hypothesis which emerged around the years 1913 to 1918 was, strictly speaking, a non-event in the annals of chemical theories. It was simply incorrect.[4] However, its formulation did serve another vital purpose. The burst of intellectual and experimental activity by international scientific communities, designed to show that reactions were not caused simply by infra-red radiation, as suggested by the radiation hypothesis, would be crucial to the next stage of progress in the understanding of chemical mechanisms. These efforts would occupy the attention of some of the finest scientific minds of the day, including Irving Langmuir, Richard C. Tolman, G.N. Lewis, and Cyril N. Hinshelwood, and would eventually benefit from the casual intervention of physicist Frederick A. Lindemann. The need to explain the failure of the radiation hypothesis and to find a feasible alternative would result in an intensified study of the effects of radiation on chemical systems. Thus the idea that reactions involving free atoms might be of significance in chemical mechanisms would come, not from Gomberg's efforts, but from studies of photochemical effects.

Nernst's chain theory of 1918 had already indicated that elementary atomic reactions could be responsible for the inexplicably high quantum yield in the photochemical synthesis of hydrogen chloride, as shown by the studies of Max Bodenstein and Walter Dux.[5] However, when H.S.

Taylor suggested in 1925 that organic reactions may occur by chain reactions involving free radicals, the idea was received as a novel one. As suggested by Nernst,[6] the primary reaction of light was the splitting of chlorine into atoms which then reacted with hydrogen molecules in the following steps:

$$Cl_2 + (h\nu) \rightarrow Cl + Cl$$
$$Cl + H_2 \rightarrow HCl + H$$
$$H + Cl_2 \rightarrow HCl + Cl.$$

This cycle would occur repeatedly, thus ensuring a very high yield of products for each quantum of light absorbed. If correct, this would explain very neatly the high quantum yield results of Bodenstein and Dux.

At the general discussion of the Faraday Society in 1925, Taylor reported his and A.L. Marshall's continuing investigations of reactions between hydrogen atoms and a number of compounds, including ethylene, carbon monoxide, oxygen, nitrous oxide, carbon dioxide, and nitrogen.[7] They proposed that 'a chain mechanism of the type familiar in the photochemical combination of hydrogen and chlorine' was feasible. Although this statement seems reasonable enough now, it must be remembered that agreement on the overall mechanism for the combination of these two elements was far from unanimous. Their conclusion in 1925 was interesting and novel: not only would such a chain mechanism explain the high yield of formaldehyde obtained from hydrogen and carbon monoxide, but Taylor also proposed that a collision between ethylene and a hydrogen atom would result in the production of a free ethyl radical:

$$C_2H_4 + H = C_2H_5.$$

This simple equation would open new fields of possible mehanisms for hydrocarbon chemistry, that vitally important area of carbon-hydrogen compounds.

The difficulties faced by free radical chemists in this period were not without irony. Even as they battled for acceptance, there were those physicists and spectroscopists especially, being innocent of the fact that free radicals were not supposed to exist, who were enjoying a much smoother passage.[8] Optical methods did not require all that clumsy chemical clutter and presented excellent opportunities for studying properties of these short-lived particles. Mere detection of free radical existence however did not solve the chemists' real problem. Spectra emitted in the excited state did not, after all, provide a quantitative measure of radical concentrations, necessary for rate measurement. Further, the degree of complexity involved in studying the progress of even the simplest of reactions proved to be formidable. Impurities, for example, which interfere dramatically with the observed rates of reaction, were major factors. Long after the pragmatic chemist had triumphantly learnt to control a chemical reaction physically and was capable of understanding *what* happens during a reaction, the *how* of the intermediate steps, or mechanisms of the process, would remain a mystery. To understand the nature of the overall reaction it was first necessary to isolate and study the individual steps which contributed to the end product. It was in this area of investigation that Steacie would make his personal mark.

As a student, Steacie was unconcerned and probably largely unaware of the problems besetting the small but determined band of scientists wrestling with questions of chemical kinetics. His attention hovered initially around the thermodynamic properties of gases, which required novel techniques and much dexterity in the manipulation of high pressure and vacuum systems quite alien to the average chemical laboratory of the day.

It was not until around 1929 that Steacie's attention turned at last to the subject of chemical kinetics. More free to pursue his own inclinations at McGill, his publications veered increasingly towards gas phase kinetics, and the study of free radicals, although a few papers on solubilities continued to appear.[9] Experimental gas kinetics would never become, in Steacie's lifetime, a neat and tidy exercise; techniques available then for measuring pressure changes, maintaining constant temperatures, and purifying gases were crude by present standards and made the control of experiments uncertain. This scientific endeavour was emphatically not for the faint hearted. But for all that, it was an exciting time to take up the subject. Steacie was now to be found, when other duties permitted, never many feet from the maze of glass tubing and vacuum lines which would serve his purpose for the coming decade. It was a time of austerity, and equipment was not all that easy to come by. In between experimental runs, he and Frank Toole would plot possible ways of acquiring another one of those vacuum pumps. But even without this addition, his output would be prodigious. Investigations began with the kinetics of the heterogeneous thermal decomposition of methyl formate in 1930. So far was the concept of free radicals from being accepted, the term itself was totally bypassed. But within a few years, there would be major changes.

Steacie searched for suitable reactions to study gas phase oxidations and the homogeneous and heterogeneous decompositions of simple molecules. By 1932, he had focused on the ethers as model systems, studying the simultaneous pyrolyses of dimethyl and diethyl ethers. The results were by no means conclusive. Earlier results which appeared to show that the rates of decomposition were additive were later shown to be far more complicated. In the four years up to 1934, he published some thirty-five papers, mostly devoted to kinetics.

Around the same time and intrigued by the same body of literature which had aroused Steacie's interest, Francis O. Rice, recently arrived at Johns Hopkins University, turned to the subject with some bold suggestions. Taylor's idea of 1925, that collision between ethylene and a hydrogen atom could result in the production of a free ethyl radical had struck an agreeable note. Was it feasible that most organic compounds, and hydrocarbons in particular, could decompose through this type of free radical mechanism? Such decompositions, Rice reasoned, could be regarded as chain reactions in which certain atoms or radicals start a cycle of activity at the end of which they are regenerated and start a new cycle. The beauty of this scheme lay in that, if correct, it could account both qualitatively and quantitatively for the otherwise puzzling result that highly complicated decompositions could, in some circumstances, show simple kinetic behaviour. Rice and Herzfeld's paper on the subject appeared in the *Journal of the American Chemical Society* in 1934. In that same year, Steacie published what was to be the first in a series of papers dealing with the unimolecular decomposition of the alkyl nitrites, beginning with gaseous methyl nitrite. Also acknowledged in this particular investigation was Rice's 'theory of free radicals,' which Steacie and his co-author G.T. Shaw adopted without further comment.

The study of the thermal decomposition of the gaseous nitrites would continue for several years with the investigation of ethyl, n-propyl, isopropyl, and n-butyl nitrites at low pressures. This homologous series proved to be an ideal system for the detailed study of reaction mechanisms with increasing molecular complexity. In all cases the primary and rate-determining step followed the pattern

$$RONO \rightarrow RO + NO.$$

The results showed that, for this series, the rates of reaction declined with decreasing pressure, in keeping with the theory of unimolecular reactions. However, there was no corresponding explanation on the existing theories for the observed rate increase with molecular weight. Steacie postulated that the relationship between chemical configuration and reactivity could be complicated by factors such as small changes in binding strengths of an ascending homologous series.

Steacie's travels in Europe in the summer of 1934 meant a break in the series of experiments but not the subject. In Frankfurt, Steacie's attention had turned to reaction rates involving heavy water, which proved to be 'just the sort of thing' he had gone in search of. From the physical chemistry laboratory at Leipzig where he found himself in January 1935, Steacie issued a long retort to criticisms by W.A. Bone of Imperial College, London. Professor Bone was displeased with several aspects and the conclusions which he had found in the paper on the oxidation of ethane published by Steacie and A.C. Plewes a year earlier. Tempering of words not yet his forte, the young and irascible Steacie spared no ground in meeting the criticisms nor his scathing remarks: 'Professor Bone accuses us of "inexperience." In this connection, all I can say is that I prefer the inexperience of youth to the dogmatism of old age.'[10]

Steacie did not allow external criticisms to impair his personal concept of what doing science entailed: hard work, long hours, tireless persistence, the best effort one could muster, always enough humour to overlook the occasional transgression, and as little ceremony as possible. When times were hard, as during the Depression, one had to show more initiative. But chemistry often provided compensations. Steacie generously passed on his recipe for punch to one grateful student at McGill, who has long since forgotten the proportions but not the components: coca-cola, ginger ale, lemon juice, sugar, and 95 per cent ethyl alcohol, the last obtainable from the departmental store-room when discreetly requested.[11]

It was at this time in the mid-thirties that Steacie's interest began to focus on a central problem in chemical kinetics: how to separate out and measure the kinetic parameters of each rate element in composite reactions. If the overall kinetics could be reduced to a stepwise series of reactions, each separately determined, their individual rate contributions to the total reaction process could be assessed. Steacie concentrated increasingly on photosensitization experiments initiated by the resonance states of heavy atoms. The process of photosensitized decompositions could be regarded as a form of predissociation of the complex formed when the reactant molecule and the excited atom collided. These reactions thus presented a close analogy to the thermal processes with which Steacie was now very familiar. Photosensitized reactions also had the advantage of often being simpler than thermal reactions. Experimentally, mercury had been most widely used, involving an excitation energy higher than the activation energy of most reactions so that it could be used to initiate a wide variety of processes. The other two metal sensitizers to which Steacie turned his attention were cadmium and zinc.

By 1938 a series of publications on investigations into photosensitization of hydrocarbons, beginning with ethane, began to appear in both national and international chemical literature. For ethane, both mercury and cadmium indicated similar mechanisms. The triplet-excited metal atom initiated an electronic energy-transfer process with the formation of the metal hydride, which could be detected by its resonance emission:

$$C_2H_6 + Hg(^3P_1) \rightarrow C_2H_5 + HgH,$$

$$C_2H_6 + Cd(^3P_1) \rightarrow C_2H_5 + CdH.$$

The primary reaction appeared to be a C–H bond split resulting in an ethyl radical. Further work with propane and butane again showed that the break of a C–H bond was essentially the primary step, followed by combination of the alkyl radicals. Thus, for propane, Steacie was able to show that using either mercury or cadmium the primary step was

$$C_3H_8 + X(^3P_1) \rightarrow C_3H_7 + XH.$$

However, secondary steps in these reactions differed depending on the metal used, a problem which Steacie and his co-workers continued to investigate.

More interesting still for Steacie was the photosensitized polymerization of ethylene, a much investigated but poorly understood reaction. Its reaction with triplet-excited mercury was generally assumed to be the formation of acetylene and molecular hydrogen accompanied by the decay of mercury to its normal state:

$$C_2H_4 + Hg(^3P_1) \rightarrow C_2H_2 + H_2 + Hg(^1S_0).$$

This would be followed by the mercury-photosensitized polymerization of acetylene together with the reaction of hydrogen atoms and of radicals with ethylene:

$$H + C_2H_4 \rightarrow C_2H_5,$$

$$C_2H_5 + C_2H_4 \rightarrow C_4H_9 \quad \text{etc.}$$

In results that would be published in 1941, Steacie suggested an alternative mechanism, when ethylene was first promoted to a higher energy level. This energy-rich molecule was then deactivated either by collision or by unimolecular decomposition:

$$C_2H_4 + Hg(^3P_1) \rightarrow C_2H_4^* + Hg(^1S_0)$$

followed by

$$C_2H_4 + C_2H_4^* \rightarrow 2C_2H_4$$

or by

$$C_2H_4^* \rightarrow C_2H_2 + H_2.$$

The investigation was continued at higher temperatures with alkyl substituted olefins. Increasingly, Steacie's work involving molecules in the excited state would lead other chemists to new research in the area of fast reactions and eventually to the field of flash photolysis.

Early in March 1939 Steacie received a letter from S.P. Eagleson, the secretary-treasurer of the National Research Council in Ottawa, inquiring whether he wished to be considered for the directorship of the chemistry division, recently vacated by G.S. Whitby. The list of candidates submitted had been growing for some months and had reached thirty-five before Council short-listed it to six, which included both Steacie and Otto Maass. McGill was very reluctant to lose Maass, and countered the NRC offer by raising his salary to $6,000.[12] By comparison, the role of director carried a salary of $6,500, and would either be permanent or for a term of years, 'depending on circumstances.' Maass decided to stay at McGill. After all, he had been appointed head of the chemistry department less than two years previously.

Steacie had been personally recommended by Whitby together with several other possibles. As a former member of the McGill faculty, Whitby was well aware of Steacie's reputation; his personal assessment ran thus:

Pro: clever, highly competent, hard working, upstanding, a first-class lecturer and speaker.
Con: No interest in industrial chemistry. Wholly devoted to making a name for himself by academic research on some of the more recent theoretical aspects of physical chemistry.[13]

If somewhat simplistic, this assessment would have pleased Steacie.

He responded with a curriculum vitae as requested by Eagleson, an introspection of his achievements to date. He was, by then, a fellow of the Royal Society of Canada and an associate editor of the international *Journal of Chemical Physics*; his teaching had covered elementary chemical engineering, colloid chemistry, thermodynamics, chemical kinetics, photochemistry, and molecular structure. Publications amounted to eighty-eight items. He saw the past decade as having been spent mostly supervising the research of graduate students (with whom he would co-author most of his publications), twenty-two of whom would have received the doctorate by the end of the year. It was worth noting that much of the financial assistance for their research programs, which covered hydrocarbon chemistry as well as kinetics and photochemistry, had come from the National Research Council. Since Steacie and his students worked in a building apart from the main chemistry faculty, there had also been a considerable amount of independent organization involved in running the laboratory. It all mounted up to good all-round experience and a remarkable record for one not long past his thirty-eighth birthday. This was no time to be self-effacing. 'I can, I think, say that I have acquired an international reputation in the fields of hydrocarbon chemistry, combustion, chemical kinetics and photochemistry,' Steacie wrote, in support of which he cited his role as editor and participation in symposia of recognized bodies such as the Faraday Society and the American Chemical Society. Industrial experience was also noted; as a consultant Steacie had worked on diverse problems, including the manufacture of dyes, insulating bricks, gas liquefying plants, radioactivity, paint, and the suppression of excessive foaming in beer. This impressive and practical list was augmented by Otto Maass, who was requested by Eagleson to provide a reference. 'My only hesitation in recommending him is the rather selfish one of not wishing to lose him from the department of chemistry at McGill ... it is perhaps superfluous to add that in my estimation he would devote his full energy to making a success of his position.'[14]

By the end of March, other candidates on the short list having replied that they did not wish to be considered, Steacie had been offered and had accepted the post and was discussing details such as pensions. But in fact much heart-searching had occurred in the interim. This move from the familiar pace of academia to a government laboratory would be a major change, not to be made lightly. On only a few other occasions in life would Steacie be equally tortured by doubt as to the correct course of action. Five years earlier he had already turned down an offer by Whitby to join his division of the NRC as assistant director, despite the fact that the post would have doubled his university salary. Academic life suited him and he was good at his profession which, in a strange way, may have accounted for the eventual decision to move on. As Maass himself pointed out, 'he has shown an unusual ability in directing the research work of his graduate students. In doing so he has at all times had their loyal support and their complete confidence.' Maass was not the only one in the department who observed that students were gravitating more and more toward Steacie's fields of interest. However, Steacie's hopes of rising through the ranks at McGill were limited; Maass was freshly appointed to his post and a long way from handing over his leadership role. Nevertheless, Steacie battled hard against the decision to leave his students and the research which he loved. Was temptation prompted now by the financial increase to double his present salary, he reasoned? This criterion alone would not have been considered; all his life Steacie would remain oblivious to the attraction of monetary rewards. More to the point, would responsibilities at a government laboratory employing large numbers of men, as compared to the eight graduate students he now watched over, mean the end of his own research – which was just becoming truly exciting? He took his dilemma to his respected mentor, Otto Maass; in the end Ottawa won.

With characteristic enthusiasm, once a bridge was crossed, Steacie told Eagleson he was looking forward to the work in Ottawa 'with the greatest pleasure' and that 'I shall do my best to justify the confidence which the Council has placed in me. This placing of confidence would prove a major factor in the events soon to engulf the NRC. By spring, the appointment had been announced in the press (where the *Ottawa Journal* made the common error of addressing him as Steacey). General McNaughton, the president of the Council, cordially told Steacie, 'I personally look forward to our collaboration in the development of the work of our division of chemistry.' Early in May, Steacie travelled to Ottawa to talk things over with McNaughton. There was now no turning back.

Parting from McGill, to which he had been attached for twenty years in one capacity or another, was not easy. Apart from anything else, students had to be provided with alternative supervisors and there was generally much winding up, to ensure that no one was left in the lurch. For those students nearing the end of their research, Steacie intended to see to their needs by commuting back to Montreal at intervals, and in the case of one particular student who held a Council studentship and was also leaving McGill to join the NRC, Steacie eased the hectic preparations of these final days by resorting to a technique he had used before. The former student, now an employee in the chemistry division to be headed by Steacie, found himself soon after with a file containing his project report to Council (required of all holders of studentships), a letter on McGill letterhead from Professor Steacie approving the report, a letter from the special Council reviewer (a chap named Steacie) also approving the report, and another letter from the director of the chemistry division to the same effect. The result was, Steacie gleefully informed him, that at least two people had been saved endless inconsequential paper work.[15] Meantime, in the summer of 1939, Steacie found himself the Director of the Chemistry Division of the National Research Council in Ottawa, at a salary somewhat higher than that of his former professor and soon to have even greater responsibilities thrust upon him.

4

The Making of an Institution: The National Research Council of Canada

The National Research Council of Canada was set up in 1916, ostensibly as the result of Britain's concern that the dominions should each have an organization akin to its own newly created Department of Scientific and Industrial Research (DSIR). It had come as a sobering shock in 1914 to the nation which gave birth to the industrial revolution that, when war called for highly qualified research men, none could be produced to speak of. In fact, 'there were more trained scientists in a few of the great German industries than could be found in the entire British Empire.'[1] On the eve of the First World War Britain had found herself still importing vast quantities of minerals, metals, and dyestuffs from Germany. Worse still, as events turned out, dependence on German technology included vital supplies for industrial needs, such as zinc, smelted from ore originating from within the empire. Even the army artillery came equipped with gunsights made by Goertz of Berlin.[2]

What Canada actually created in response to the British request was a committee of cabinet ministers – the Privy Council Committee on Scientific and Industrial Research (PCCSIR) – which in turn appointed an Honorary Advisory Council or National Research Council (NRC) to be caretaker of the nation's efforts in these endeavours. Some confusion would arise later from the fact that the research laboratories with which the name NRC is now synonymous did not materialize during the first sixteen years of the Council's existence. Their establishment was finally achieved in the midst of the Depression by dint of one Henry Marshall Tory.

The creation of the advisory body was warmly welcomed by universities across the country and the Canadian Manufacturers' Association, both of which provided men, eleven in all, to serve on the Council as honorary members. The Council was charged with 'the promotion of researches which would develop industrial production and the utilization of the natural resources of Canada.' How to implement this lofty order was another matter. All members concurred, however, that Canada's immediate and future development in both resources and industry would require a more solid scientific foundation than then existed. The Council's proposed mandate was thus as broad as it was ambitious; from the beginning the 'Council took itself pretty seriously.'[3] Under the chairmanship of Dr A.B. Macallum, professor of biochemistry at the University of Toronto, the men saw the tasks ahead as vital to the nation's development, a major undertaking entrusted to their care.

One of the Council's first tasks was to survey the state of scientific affairs in the country, starting with industry. Out of the Dominion's ten thousand industrial firms which were approached, and of the twenty-four hundred firms which responded, the figure then spent per year on industrial research appears to have amounted to just $135,000 (salaries not included). Only two universities carried out extensive research and, in the twenty-three years up to 1917, only eleven doctoral degrees in science had been granted in the entire country. Macallum subsequently estimated that the whole of Canada at that time possessed less that fifty persons engaged in fundamental scientific research.[4] Things did not look promising. To the men assembled for the purpose of co-ordinating science in industry throughout the land it was a stark realization that industrial research in Canada was virtually non-existent, a situation matched only by the lack of manpower to undertake the work. There was, in effect, very little to co-ordinate.

The problem was certainly not new. In 1882 Sir William Dawson, the principal of McGill, lamented that the government was making no provision for scientific research, apart from some meagre grants to the geological survey and meteorological services. 'We have no national society or association comparable with those in other countries,' Sir William had observed, 'yet we are looking forward to a great future.'[5] Now that such a body had actually been created there was much to do and no time to be lost. The resources available, however, contrasted miserably with the immensity of the problem. No central office existed; no clear budget had been assigned. The entire staff consisted of one busy civil servant with priority obligations elsewhere. In fact, all the members of Council were simultaneously occupied with other full-time posts and no arrangements had been made even to reimburse their travelling expenses.[6]

Physical discomforts aside, the Council members were determined and enthusiastic in the face of a novel challenge. Very quickly, however, a number of fundamental issues became apparent which were not related to the job of co-ordinating science in the country, but to the very nature of the institution itself. If it were to survive, a few basic things would have to be made manifestly clear. To begin with, what was to be the status of the Council within the government hierarchy? It had been created by order-in-council; it could well be abolished in the same way. Even as Council debated this problem, plenty of others arose which vied for attention. To give meaning and substance to all the plans under consideration, it was imperative to have a base from which to operate. As early as 1917 the Council proposed that 'there should be a Central Research Laboratory and Bureau of Standards at Ottawa for fundamental problems.'[7] Its purpose would be the pursuit of both pure science and 'science applied to the industries in Canada.'[8]

Now was an appropriate time to cast a glance across the border. There, already outstandingly successful in the American industrial drive, was Pittsburgh's Mellon Institute. The brain-child of Robert Kennedy Duncan and the philanthropic gift of the brothers Andrew and Richard Mellon, the institute regarded itself as 'a guild of scientists established for the promotion of comprehensive pure and applied research on important problems.'[9] Established in 1913, its strength had grown by leaps and bounds. Within a decade it was rumoured to be spending $20 million annually in research between the three major areas of chemistry, physics, and mechanics.

Duncan was a Canadian, born in Brantford, Ontario, in 1868. Addicted to science at an early age, he had continued to graduate work in chemistry at Columbia University after graduating from Toronto with a BA in 1892. Although immensely interested in the rewards of research all his life, Duncan was manifestly a proselytizer of the benefits of science. During his short life the young enthusiast laboured incessantly under the 'intense conviction that only through the broad application of modern science to industry will there ever come into the world an era of gracious living.'[10] After touring Europe and surveying the wonders of radioactivity in Pierre and Marie Curie's laboratory in Paris, Duncan conceived the idea of his industrial fellowships, having observed the strength derived from co-operation between German industry and universities. The speed with which the European centre of scientific excellence had passed to Germany in the nineteenth century had not gone unnoticed, nor had the methods by which it had been achieved. The German Chemical Society, for example, was publishing an index to the world's chemical literature as early as 1830, whilst the British Chemical Abstracts date from 1882 and the American Chemical Abstracts had not begun until 1907.

It was a bitter pill to know that Duncan's ingenuity for motivating national resources to human profit should have had to seek an outlet in the United States. For Canadians interested in higher education, Duncan's fate was not unique at that time though his achievements probably were. From the beginning, the members of Council, called to appear before the Cronyn Committee set up in 1919 to debate the whole scientific research problem in Canada, stressed loudly the exodus of existing scientific talents to the United States. Bolstering that nation's universities and industry was accompanied by a matching loss for Canada. This was reason enough for urgent action. DSIR notwithstanding, what was being proposed now for Ottawa was an institution which combined the roles of the National Bureau of Standards (another equally successful American institution established in 1901, based in Washington, DC) and the Mellon enterprise; two individual organizations, united by excellent work. The Council would create for Canada a new institution possessing the merits and characteristics of both. In Steacie's lifetime, the NRC would receive high accolades and a compliment from the Mellon organization, a model which, in the beginning, it had only hoped to emulate.

The other contentious issue raised at the committee hearing and well taken by members of the Council related to the difficulties of practising science under the Civil Service Act. A later observer of the scenario may well contemplate the high probability, indeed the 'inevitability of tension between an organization structured and motivated like the Civil Service Commission, and the Research Council.'[11] This was due to the curious interplay of intentions, both similar and in antithesis, which marked these two bodies. The commission, created in 1908 but still far from free of internal problems a decade later, saw its mission as that of a central control agency. This control ethos already lay heavily on those government departments under its jurisdiction when the Council's mandate was being debated. If Council took its task seriously, the Civil Service Commission was no less zealous.

At the time of the Cronyn inquiry, scientists working under federal jurisdiction found an opportunity to air grievances with some passion. 'Government control,' Charles Saunders, the Dominion cerealist, pointed out, 'is based on the idea that immediate success is to be aimed at, no matter what sacrifice of ultimate good may be necessary.' This was a direct corollary to the fact that all governments aim to please the people; after all, they hoped to be re-elected. The drawback was that the ordinary voter wanted results right away. Still, governments come and governments go. Much more pertinent to the present inquiry was the role of the Civil Service Commission itself, a permanent fixture, and the question of how the new institution would be organized. Macallum voiced his feelings strongly. 'If [the proposed National Research Institute] is put under the Civil Service Commission it will die prematurely or be ineffective as so many Departments of the Government are.'[12] This harsh view of bureaucratic efficiency no doubt stemmed from long-held prejudices against the permanent government administrative staff. As for external control, science by its nature was not given to such whims. In short, scientists could not create to order or do good work when thus regulated. A scientist busy thinking should be left alone and allowed to think for as long as needed. The idea of having to sign an attendance book at nine o'clock next morning must not be permitted to interfere with a scientist sitting up all night should work require it. If such freedom was not available, 'his work will seldom amount to much,' Saunders had told the listening members of Parliament.[13] Everything considered, the Council wished earnestly that the Institute should be placed under its own control. 'It would be better not to establish such an Institute than have it controlled by a Department ... The Research Council ... should be responsible not to any one Department but to the Cabinet as a whole. This is the situation in England, where the Department of Scientific and Industrial Research ... is responsible to the Privy Council only.'[14] Thus, in a spirit of scientific autonomy, were the two major tenets of the Research Council formulated: to estrange itself from control by any central agency, and to counteract the sting of 'foreign superiority.'

The early years of the Council's existence were not to be smooth sailing. By 1921 its plans were a shambles, its proposed legislation defeated in the Senate. Another two years passed in which successive chairmen resigned in chagrin, frustrated by the government's lack of action. In April 1923, about the time when Council members' spirits had reached a nadir, a new member, H.M. Tory, the president of the young University of Alberta, joined the Council; by the end of that year Tory was its new chairman and in 1928 became the Council's first president.

The advantage that Tory possessed over the NRC's previous leaders was his long experience in the art of scientific persuasion; that is to say, he had worked out the most logical way to counter antagonisms and the best possible way to put over the reasons to and for science. In fact, by the time he joined the Council, Tory had already tasted plenty of the kind of objections popularly raised at the idea of scientific research. As the chairman of a meeting in Alberta, which had been addressed by Tory, so eloquently put it, 'We have listened with pleasure to the Doctor but I want to tell you people that God made the world and I suggest he knew what he was doing and the good Doctor can't change it and even if he could, what would we do with all the stuff we would grow if we followed his advice?'[15]

From such an onslaught a lesser mortal could be forgiven for retiring, vanquished. Tory's advocacy for science was, however, just beginning. This inexhaustible zeal owed something perhaps to the fact that his own conversion to science from the originally intended religious studies had come about so unexpectedly. Besides, his brief sojourn to Cambridge in 1892 had shown him a world outside Canada's pioneering communities. Cambridge had itself not long awakened from a period of relative scientific sloth and entered a new age with a superb school in physics which, having entertained Maxwell and Rayleigh as its first Cavendish Professors, now boasted scientific giants such as J.J. Thomson and Sir Gabriel Stokes. Tory would also have learnt that this apparent esteem for science at Cambridge (and Oxford for that matter) had not been easily won, nor had it been enjoyed for long. Its pioneers had fought a long and hard battle against a rigid university structure and archaic traditions. Undoubtedly this knowledge strengthened Tory's resolve in the face of future adversities.

By the opening years of the new century, Canada itself was hastening to establish new centres of learning. But existing colleges in the east were still obliged for the most part to send their graduate students to Europe and the United States because of the lack of facilities at home. Tory, now on the staff of McGill, tried to raise the status of his university to include a graduate school, which came into being in 1907. This was to be just the first of many Tory innovations. From 1908 to 1923 he would lend his talents to the creation of several new institutions devoted to the promotion of knowledge and other heuristic ends. Even the NRC was not to be his final contribution to the art of 'pioneering.' On leaving the NRC in 1935 there was time and energy to found Ottawa's Carleton University.

If the NRC was neither Tory's first nor last undertaking, it was by far the most significant for Canada as a whole. The coming together of one such as Tory and an embryonic institution, struggling now for existence, was indeed opportune. It was a task much to his liking, made for his pioneering spirit. There was little time to be lost; if the quashed plans for laboratories were to be revived, action would have to speak louder than words. For Tory, the next nine years would see his drive to fulfil the Council's ambitions fully tested. He initiated a survey of the major problems facing industry around the country and, through the Council's associate committees, organized co-operative research on urgent national problems. All the while, Tory continued to address scores of public meetings and to generate interest in the press.

The results were slow to come, for diverse reasons. Partly the advancement of scientific research in Canada was not viewed as a national priority. In Parliament, one member saw government grants to universities for research as a sinister precedence. 'Men thus engaged would consider themselves pensioners of the state and would idle away their time.'[16] Some of the reasoning had a fateful overtone. As Arthur Meighen, then leader of the opposition, put it in 1924, 'We have to keep abreast of scientific advance, but we cannot afford to go into work on as big a scale as other countries. We get the advantage, it must be remembered, of what other countries do, but chiefly we get the advantage of United States activities, because their scientific problems are virtually the same as ours, especially in agriculture.'[17] This sentiment of dependence on foreign powers would take on new meaning sooner than anticipated.

The particular agricultural problem to which Meighen alluded was that of wheat rust, a serious disease affecting cereal crops. Losses to the country in bad years were monumental, estimated in 1927 to be $100 million. It was a national problem on which Council focused attention early and with phenomenally successful results. This success story permitted co-operation between the Department of Agriculture, the universities, and the Council.[18] Savings to the nation would be enormous.

Curiously, in light of subsequent events, the NRC suffered in its early years from a lack of support by the academic communities at large, perhaps for fear of competition. Inevitably, there were also debates on the location of the new research facilities. Why should Ottawa have all the benefits? The president of the Empire Club of Toronto proclaimed his city to be 'one of the greatest industrial cities of Canada, and indeed of the Empire at large'; if there were going to be research laboratories (not to mention an information bureau and a technological library) they should be situated, in his opinion, in Toronto. As would occur more frequently in the years ahead, the press and public opinion finally accelerated events. Tory saw a flicker of light at the end of the tunnel when the Toronto *Globe* reproachfully addressed the country: 'We are lamenting the departure of university trained men from Canada. Many of them have been drafted by countries spending millions of dollars annually on research ... the field awaits them at home to their own profits and that of their country, but we do not provide it.' And while the rust problem cost western farmers millions of dollars, Parliament had only seen fit to pay out $2,700 in salary for one trained pathologist to attack the problem. Britain was spending millions of pounds on scientific research, 'but Canada is hopelessly immovable.'[19] Such persistence, persuasion, and propaganda as could be mustered throughout the country by the Council and Tory were helped by the gradual lifting of economic and political pressures.

In August 1932 the laboratories of the National Research Council were officially opened by the governor general, the Earl of Bessborough. Prime Minister R.B. Bennett, who had supported the idea of national research while in opposition but was perhaps uncertain what scientific research would hold for the future, thought it best to be general. 'The purpose of this building,' he told the gathering, 'is to determine how industry and mankind can best be served.' The building thus dedicated was a splendid edifice, set in ten acres looking across the Ottawa River, its entrance framed by eight Doric columns. For a cost of around $3 million it provided some 270,000 square feet of laboratories and office space. A grand staircase, curved in the Italian Renaissance style, two exhibition halls with vaulted ceilings located beneath the interior courts which would display the pride of scientific and industrial progress, and a fine auditorium for scientific meetings fully equipped with a preparation laboratory added to the special features – 'utility and beauty' happily combined. The panelled board room had an ornamental ceiling, distinctively Georgian in style, while down the corridor the ceiling of the president's room was adorned with figures in low relief, appropriately depicting four areas of science. A high chamber, rising to a beamed ceiling, housed the library and reading-room. The laboratories were plain and functional, with floors of a special magnesite composition developed by the Council's own staff.[20] It was a building unique in Canada. The past and the future, the sciences and architectural heritage combined, the best of what Tory had observed in research organizations in Britain, France, Germany, and Japan, had been brought together under one great roof. It had been a long battle but the future looked all the more hopeful for it.

Ironically, these days of final triumph were not to be happy ones for Tory personally. It was his misfortune to have made an early enemy of the man who was now prime minister. Corbett, in his biography of Tory, is blunt: 'It had taken R.B. Bennett twenty-five years to catch up with "that man Tory" as he was wont to refer to him. He waited a long time to avenge himself upon the man who had defeated him too often and too thoroughly in Alberta.'[21] Tory's official ties to the NRC and to the laboratories which he had guided and brought into being were to end painfully and somewhat abruptly in the summer of 1935.

The man chosen by Bennett to succeed Tory was as unexpected (by the Council staff) as the choice was sudden, although not as abrupt as Corbett and later commentators have implied.[22] Major-General A.G.L. McNaughton was an electrical engineer by training but up until this time had been a professional soldier. This switch from chief of general staff to chief of the NRC was desired by him no more than by members of the young institution. Not surprisingly, both parties were at first less than satisfied; each eyed the other with some dismay. This mutual consternation had some basis, for the habits of twenty years die hard. McNaughton, in the early phases as leader at NRC, found adjusting manifestly difficult and 'endeavoured to introduce a very formal military procedure.' It was an innovation doomed to failure, as scientific spirits could be guaranteed to eschew army protocol. The small staff at the laboratory had been led from the beginning by the gentle and scholarly Tory, who was by then 'worshipped by many of the young scientists and loved by some of his old cronies.'[23] Now they found themselves facing this handsome but sternly commanding general who was in the habit of striding through the corridors, 'his coat tails trailing in the slipstream he created as though they found it difficult to keep up with him.' More than one speculated that these were shoulders more fashioned for epaulets than a laboratory coat, more accustomed to taking command than delegating responsibilities.[24]

Fortunately, all this proved temporary, a superficial manner which 'obscured the true McNaughton.' The general learned quickly not to intrude upon his staff's liberty and, as C.J. Mackenzie recalled, 'soon the people just worshipped him.' As the man destined to inherit the whole operation, Mackenzie's opinion is more relevant than most.

As matters stood early in 1936, it could hardly be said that the NRC was in a particularly healthy state. The Council by this time had been in existence for twenty years and could look back on an unimpressive birth, on many discouraging early years, and at the depression and malnutrition of the five or so years just past. McNaughton himself was distressed by his first impressions of the Council's headquarters on Sussex Drive. Outwardly, the structure itself stood solid and confident, braced to meet the future. Inside, scientists and technicians found themselves working in bare rooms with little equipment.

Fortuitously for McNaughton and the Council, the years of economic crisis were coming to an end. While the Depression lifted only slowly and financial problems at the Council would not disappear overnight, things were changing. McNaughton took his case all the way to the top – to the man who had put him in the seat, Prime Minister Bennett: 'I stood up there in front of the [Cabinet] Council ... I think I made the most impassioned plea for scientific research ... that I have ever made.'[25] McNaughton had won his case and these were still early days in office.

The organization of a research establishment was and would remain notoriously complex. In the days before management became an art and brought with it irrevocable changes, the success or failure of an institution rested heavily on the abilities and philosophies of its leader. His task was to juggle skilfully opposing demands. He had to have clear objectives on which particular work to develop and which not, and select the kind of staff who could be stimulated to ensure productive output. He must impose a certain amount of bureaucratic organization to keep the institution running smoothly, yet not so much that it would kill originality. He must possess the ability to predict future trends while holding present policies, all the while adjudicating between the relentless competition for scarce funds. And simultaneously with all this, he must please the government which had to answer to the public.

How, when, where, and how much to spend carrying out these diverse objectives would frequently engender invidious decisions. Besides, how could there be communication and understanding on something as abstruse as scientific research when no one yet knew the rules? For one thing, the whole concept of a research institution was a recent invention, new to most countries, given impetus by war as much as anything else. For another, the activity itself, 'organized research,' was still being tested. Was research, in fact, organizable? Of course there were the few outstanding examples of early research groups: the Cavendish Laboratory, then ruled by Rutherford, or the Physikalisch-Technische Reichsanstalt in Germany, established in 1887 under the illustrious Helmholtz, which had spawned the National Physical Laboratory (NPL) in England a decade later. As government-sponsored scientific establishments, the NPL and the National Bureau of Standards in Washington were to have many things in common – including problems – with the NRC. There were also the highly successful American industrial research laboratories. The General Electric Company (GEC) founded in 1901, the Bell Laboratories which gained autonomy in 1925, and the Mellon Institute. One advantage of a later start is the opportunity to watch and analyse the progress of predecessors. Tory, as a major part of establishing NRC priorities, had toured Europe and the United States and visited the NPL, the Kaiser Wilhelm Institute at Dahlem, the Bureau of Standards in Washington, and the Mellon

Institute. It was a thorough and highly useful exercise, from which much necessary wisdom was gleaned. Even so, each of these organizations encompassed only a part of what NRC was expecting to accomplish, even though their individual budgets were often much greater.

McNaughton reflected upon many of these problems early in his tenure. As he saw it, the business of research was 'to serve an expanding civilization, to find new and better methods of satisfying old requirements, to open the way to new services needed by the public, and to create the materials and techniques through which they may be developed.'[26] It was a truism throughout the last century that industries everywhere had been founded on the basis of scientific knowledge built up over many centuries. Mechanical industries derived from Newton's laws of motion; electrical industries were dependent upon the discoveries of Henry, Faraday, Maxwell, and Ampère. Chemical industries were the modern progeny of ancient knowledge accumulated from days when chemistry was still an art. This fundamental knowledge had taken a long time to store up; it had been acquired on the whole not for utilitarian purposes but as a by-product in the search for truth. Few of these discoveries had found any application for the decade in which they were revealed, but for the days and years ahead the results of the accumulated knowledge would alter the world in a phenomenal way. Even Edison, that prodigious genius, had devoured Faraday's work on electricity in his youth only much later to give birth to the electrical industry. By the beginning of this century, industrial application had absorbed such a multitude of these scientific ideas that new facts were now required; thus had the need for organized industrial and scientific research been born. As for industrial research by private industries in Canada, precious little existed. Increased efficiency in industry, McNaughton repeated, was the major aim of the NRC.

By 1937, when McNaughton was expressing his views to the government, the General Electric Company, Bell Laboratories, and the Mellon Institute were already deeply engrossed in both the pure and applied aspects of research. Around this time, W.A. Hamor, assistant director of the Mellon Institute, and his colleague E.R. Weidlein, took a glance at industrial research and pointed to some of their own areas of achievement. In addition, they listed a number of the most notable advances of science from 1914 to their day, 1935. In 1914 panchromatic motion picture film was marketed by the Eastman Kodak Company, Corning Glass Works introduced Pyrex glass, Moses Gomberg was rewarded for his contributions to the study of the anomalous free radical, and T.W. Richards was honoured as the first American Nobel laureate in chemistry.

In 1915 dry-lime sulphur insecticides were developed, and Irving Langmuir of GEC was lauded for his research on chemical reactions at low pressures. In 1916 x-ray methods were applied to industry by GEC. In 1924 catalytic oxidation of ammonia was applied extensively to manufacture nitric acid, and G.N. Lewis was honoured for studies on the valence and stability of atoms. In 1932 halogen methane derivatives were developed as a refrigerant and Irving Langmuir received the Nobel Prize for research in chemistry. In 1935 a new type of rayon for tire construction was described by Dupont, tempered glass with sufficient strength, when clamped to the end of a diving-board, to support a weight of 110 pounds was marketed, and J.A. Nieuwland was an American Chemical Society medallist for his basic researches on unsaturated hydrocarbons. The list of achievements through the passing years is undeniably impressive.

But what was most revealing in this survey of twenty years of American scientific highlights was the simultaneous achievement of industry and the development of pure research, pursued for no direct 'utilitarian purpose,' as McNaughton had expressed it. Both were recognized and rewarded. There was, in this period, simultaneous growth of academic as well as industrial laboratories, the establishment of new industries, and of academic journals and societies to promote their special interests. This parallel pacing of development on both the pure and applied fronts would make the United States the most technologically advanced nation in the world in a relatively short time span. As its nearest neighbour, Canada would surely be affected.

Meanwhile in *its* first twenty years or so of existence, the NRC had concerned itself chiefly with domestic matters, and the nation's natural resources. Witness, in the year 1934–35: studies on the temperature, relative humidity, and air movement in celery or poultry storage rooms; determination of the freezing point of canned fruit; cleaning and handling of barley; weather and wheat yield in western Canada. Witness, in the division of chemistry: studies on Alberta bitumen, on sugar-beet table syrup, the properties of honey, industrial utilization of potatoes, and on laundering and dry-cleaning. Witness the miscellaneous investigation of adhesive for postage stamps, and floor detergents; and in physics, the study of aircraft skis and farm windmills. There were also a few more exotic items: for example, plant hormones, and the properties of radium compounds. Unbeknown to those involved, some of this early work on x-rays and radium would have far-reaching implications.

In the mid-1930s the position of industrial research in Canada was similar to the situation in other modern industrial countries, but with an added element. Many Canadian industries were merely branch plants of larger corporations based in the United States or Europe, and had relied on their parent organization to provide short-term solutions to industrial problems. In McNaughton's view, the country was now paying tribute on an enormous scale. The solutions to this problem, if any general ones were to be found, would not be easy to implement. What was called for was a sure and steady growth of indigenous science, a process which would most assuredly require time.

McNaughton's perspective of the future was marked by his past. On the battlefield in the war of 1918 he is reputed to have greeted the news of the Armistice with some annoyance. 'Bloody fools ... we have them on the run. That means we shall have to do it all over again in another twenty-five years.' Whether this was an apocryphal tale or not, in 1935 a few Canadians saw the immediate years ahead in a different light from most of the population, who only wanted to get on with their lives and forget about another war in Europe. McNaughton was personally convinced that war was both impending and inevitable. While still chief of general staff, he had endeavoured continually to tell the public of the vital changes in the armed forces: 'We have entered on one of those phases of history marked by the most intense application of human knowledge to the development of machines for war,' the general wrote in 1934. 'In history we are not long concerned with nations unable or unwilling to keep pace with armament development.'[27] From being a leader of those engaged in using the machines of war, he now found himself at the NRC, in the business of national research. His mind focused constantly on the prospect of the NRC at war.

This penchant for defence research in the absence of evident hostility did not go unnoticed in government circles. 'I have done my best with my colleagues,' Mackenzie King later wrote, 'to remove prejudice which I know there has been against him [McNaughton] on account of his tendency to organize matters to the maximum with respect to possible conflict.'[28] In 1938 the general strode into the NRC's purchasing office and asked that they make haste to get microscopes which had been ordered out of Austria before war was declared. As for those bewildered by which war McNaughton was alluding to, enlightenment would not be long in coming.

It was now nearing the end of McNaughton's era at the NRC; in keeping with the times, a final touch was added, completing the Council's solemn role as keeper of science for Canada. Early in 1930, Tory had invited the prime minister, then Mackenzie King, to suggest some words of wisdom to be inscribed on the building. The prime minister had taken the task seriously and eventually chose a quotation from the first and second books of Ezra of the Apocrypha, though he changed the order of phrasing.

Great is truth, and mighty above all things: it endureth, and is always strong: it liveth and conquereth forevermore. The more thou searchest, the more thou shalt marvel.

Now prime minister again in 1938, King raised the subject and justified his choice to the minister of public works. With regards to truth, he wrote, 'it seems to me that you have the first important note in all scientific research.' The remainder of the passage signified 'the secret of progress which depends more on scientific research than on aught else.' King had indeed given much thought to the matter. 'You will see,' he told the minister, that the words combined 'thoughts of the material universe and the universe of spirit, each being discoverable only to research.'[29] The minister could not dispute that. Council had already debated the inscription back in 1931 and approved; seven years had passed and yet no action had been taken. So it came about that in the spring of 1938, the verses chosen by King were finally engraved into the stonework high above the entrance to the NRC laboratories.

Between 1916 and 1939 the NRC had been led by six different men of diverse characters and for varying amounts of time. From an initial budget of $91,600, and one full-time staff,[30] expenditure had now grown to a figure just under a million dollars and a staff of three hundred. From its first cramped and temporary laboratory on the top two floors of a commercial building on Queen Street, it now occupied a building of grandeur on Sussex Drive. Council's final mandate embraced nothing less than the realms of innovation, co-ordination, and development; the NRC was to have 'charge of all matters affecting scientific and industrial research in Canada which may be assigned to it by the Committee' of the Privy Council.[31] The time had come to be put to the test. As if to oblige, war came to call; the NRC responded in just the manner the founders intended. What events lay ahead for the NRC, for its staff, and for the country would form a special chapter of history, at home and abroad. Notwithstanding the changes which were bound to come, a few perplexities would remain; the same problems surrounding the birth pangs of the NRC would continue as critical issues in the years of its growth. The lack of indigenous research at universities, the weakness of long-term industrial research, the degree of government control in the national laboratories, and the role of science within a government organization – these would dominate the minds of future policy-makers, as they had in the past. In the post-war era a long-held objective would finally become reality: national scientific autonomy. The direction along this path was faintly pointed out by McNaughton's immediate successor, but only traced out clearly during Steacie's days. Nevertheless, the old dilemmas remained. Given a brief respite during the united efforts of war and the prosperous years of reconstruction which followed, the same debates would raise their voices with a vengeance in the closing years of the fifties and the decades after.

5
Progress through Problems

When General McNaughton became president of the National Research Council in 1935, few could have foreseen that it would not be he but another who would guide the Council's scientific efforts during the turbulent years of war. This man was Chalmers Jack Mackenzie, more often addressed as Dean Mackenzie or C.J., then dean of engineering at the University of Saskatchewan and a graduate in civil engineering from Dalhousie and Harvard universities. For five years, C.J. Mackenzie was to carry out his role as acting president with remarkable diplomacy and conspicuous success; subsequently as president he would prepare the ground for a new era of science in Canada.[1] McNaughton returned to his military career as inspector general of the First Canadian Division only a month after war was declared in September 1939. His instructions on leaving the NRC in Mackenzie's care were unambiguous. At the precise minute of the appointed hour, 'the General stood up, handed me the keys of his office, and said "I am going into uniform this afternoon; the command is now yours. We will now go to lunch."'[2] Lunch was duly taken at the Rideau Club, that august Ottawa institution which would play an increasingly prominent, though always discreet, role in statesmen's mode of communication. On this particular occasion, the food was incidental, but the location was obviously not. In years to come, Mackenzie would see this gesture as having paved the way and opened doors for immediate personal contact with the most senior people of the day, thus expediting greatly the serious business of preparing science for war. As the incumbent to an office 'very different from any other post in Canada,' the occasion was a first meeting, as a friend, with the chiefs of staff, the military missions, and diverse ministers of state. This short-circuit route for establishing credibility was, in C.J.'s opinion, one reason for the happy outcome of his appointment. Because of it, 'there wasn't any break. It allowed us to go straight along and I had the confidence, as a stranger of all your [McNaughton's] friends. It gave us a year's start.'[3] Since time is of the essence in war, this boded well.

Mackenzie was not new to the NRC. He and McNaughton had been appointed to the Council on the same day in 1935 and by the same minister; he had also served as chairman of the Review Committee, which, as the name implies, reviews the work of divisions and reports findings to Council. The experience had given him a good overview of how the NRC functioned; the knowledge would serve him well. From the beginning, C.J. did not so much view his new responsibilities as a command, more a holding of reins, to channel and guide his men (for there were few women). In the next two dozen years and more, C.J.'s philosophy and efforts, combined with those of his successor, Steacie, who shared both, would create a scientific institution without equal in Canada.

Steacie and C.J. Mackenzie took up their new positions at the NRC in the same year, 1939, within months of each other. This early acquaintance also boded well; empathy and understanding would be much needed in the times ahead. No record appears to exist of their first meeting, but Mackenzie soon arranged an informal gathering of division directors. For the present, there was advantage in being a relatively small organization; the heads of divisions attending amounted to three: engineering, biology, and chemistry. Physics was not represented as the director, Dr Boyle, was in England at the time. Perhaps with the memory of General McNaughton's first council meeting still fresh in his mind ('to say the first encounter was formally correct would be an understatement,' C.J. later recalled), Mackenzie was deliberately informal. But he had made notes for the occasion which constituted a guide to the running of the NRC that, as it turned out, would be perpetuated long after his own term of office was over. The purpose of the meeting was to put over to the directors his personal attitude to administrative policy and to learn theirs; the success of NRC, he noted, would depend upon the research divisions. Administration was necessary but it existed only so that laboratory divisions could operate at maximum efficiency. Of those present, Robert Newton, J.H. Parkin, and Steacie, none would have wished to argue with that. 'I must have your genuine cooperation, you must have my confidence and support,' C.J. told them. 'I can't guarantee perfection, my decision must be final and often I can't disclose all the factors but if we assume good intentions and good will on both sides our difficulties will be minimized.'

War was serious business and harmony was imperative: 'I want and need your loyal and active cooperation,' C.J. told this small group. Since the occasion was also a prelude to preparations for war, they discussed the matter of security.[4]

This brief meeting was the first of many. But for the men present then, the style of their new leader was demonstrated with remarkable clarity. C.J. was not one to stand on rank, nor was he above treating his staff as equals and asking for their support and advice. What he would receive in return was a great deal of respect. In the years ahead, he would surround himself with capable advisers, pay heed to their advice whenever conditions permitted, and always give due credit. With time, he became an expert in acquiring the most distinguished scientists available and nurtured these talents by encouraging creative energies to be spent in the pursuit of good science – pure or applied. C.J. was, as Thistle has observed, notably people-oriented: 'the part of nature that interested him was human nature.'[5] Mackenzie was gifted with an uncanny ability to comprehend and analyse the thoughts and attitudes of those with whom he was associated.

Steacie very soon came to understand the line his leader would take, and it suited him well. His brief term under General McNaughton's direction had not been as smooth as either might have hoped after the initial friendly exchange of letters. Steacie found the constrictions of working in a government laboratory manifestly different from life as a university don. Only recently, he had had a confrontation with the general. It happened that one of the staff, F.E. Lathe, could not reconcile his differences with the general and had asked Steacie to give him laboratory space, which Steacie did. When McNaughton attempted to step in, he was met with unexpectedly fierce opposition. As Robert Newton remembered it thirty-five years later, 'Steacie went right on the carpet with McNaughton and said, now is this man in my division or is he not? If he is in my division I am going to decide what he should do, if he's not then I'll take no responsibility. McNaughton backed down and said, Oh yes he's in your division. I won't say another word, so that was that, you see.'[6] After all, Steacie had left the military life as a youth because it engendered a discipline of the spirit which made him uncomfortable; he had not changed his mind in the decades since. For C.J., getting to know his director of chemistry would take a little longer, though in that small community, Steacie's reputation at McGill was probably known to him.

Steacie began his duties as director of the chemistry division that summer of 1939 with a thorough tour of all the chemical laboratories and staff now under his jurisdiction. He took a while to do it, but the decisions he came to were to have lasting implications. As he saw it, much of the work going on appeared haphazard, with small isolated knots of people having no overall strategy. Leo Marion, almost the same age as the new director but already a long-time member of the chemistry division, remembered Steacie coming round to nose out the situation. They had been acquainted since McGill days, though perhaps not as well as they were about to be. Sitting on the heating coils (it was summer and therefore there was no heat) Steacie heard Marion's views. There had been fights with Whitby who always wanted to take Marion away from his organic research 'to work on some damned applied thing.' Marion's dedicated research into the biosynthesis of alkaloids was showing promise and it required total dedication and commitment. Steacie made a decision on the spot; he told Marion he would never have to write another report – 'that's finished and I'll never ask you to work on something else, never.' It was a first demonstration of Steacie's brand of leadership. Marion, in retrospect, was quick to point out that if at any time his research showed lack of results, it would have been stopped.[7] Steacie's method of doing things, however, would not change. He mapped out a plan which he would soon put to the Review Committee.

That fall, Steacie found himself about to become deeply immersed with eventualities of war. However, he had no intention of allowing the administrative duties of the division to dominate his time. Initiation into NRC's working policies nevertheless came quickly: in the second week of September McNaughton pointed out at the Review Committee meeting, to which Steacie had just been invited, that investigations in the chemistry division hitherto had been directed toward utilization of Canadian natural resources and improvements in manufacturing. Of course, now that war had been declared, things would change. This was Steacie's earliest opportunity to test the water: however McNaughton saw 'our division,' Steacie clearly outlined on this occasion *his* own ideas for running chemistry at NRC. Reorganization, he told the committee, was under way so that members would work in groups rather than individually, based on an area of chemistry, with a senior man to supervise the unit, and one permanent assistant to the group leader. Something ought also to be done to improve salaries, he told the meeting. For his own part, Steacie had already appointed F.G. Green as administrative assistant to deal with routine problems of the division, in the hopes of 'leaving the director free to concentrate on research activities.' These were the interesting series of researches begun at McGill; if the Council agreed, he proposed to continue the active pursuit of these researches. If Council was surprised in any way, it was not recorded, but deemed 'this would be extremely desirable.'[8]

Still, however desirable, other considerations had to be dealt with first, as Steacie soon discovered. The NRC was in the business of mobilizing for war, and since the summer a steering committee had been formed that would soon give birth to the War Research Committee. The problems to be dealt with were multitudinous, with research priorities and the co-ordination of personnel and facilities heading the list. By the end of the year Steacie was launched irrevocably into the world of committees: the Review Committee, the Advisory Committee of Industrial Chemists, the Subcommittee on Chemical Problems, not to mention the Committee on Laundry Research. The list grew as the days passed. Never again would he be able to divorce himself from the peculiar rituals of committees, nor would he ever grow accustomed to them.

Happily for Steacie, many of these committees on which he now found himself had in the chair his former mentor Otto Maass, whose presence helped to bridge the past and present. As Steacie knew well, no one in calmer days epitomized more the dedication of the research scientist. Now, with the onset of war, no one was more zealous than Maass, content to leave the ordered world of teaching to do battle in the very different process of organizing science in war. Of all scientists in Canada, no one had been more adamant than Maass that the country had to prepare itself, and that science had a role to play. When Mackenzie was appointed acting president in October, one of his first callers was Otto Maass, who was terribly upset that the conference of industrial chemists, held earlier in the summer, had produced no action. 'He wanted something done and he wanted it done immediately.' C.J. responded by appointing him to the NRC and giving him a job to do at once.[9] It was to be one of the first of many inspired appointments, although the realization of the wisdom did not transpire until sometime later.

Maass was no ordinary patriotic citizen. His efforts at McGill had been crucial in getting graduate research established in the chemistry department, and many students had thus been launched upon their scientific careers. Now, at the outbreak of the war in Europe, O.M., as he was often addressed, found himself director of the newly formed unit on chemical warfare and smoke, under the auspices of the master general of the ordinance. This unit became central in organizing the development and production of war gases, shells, smoke weapons, and defensive chemical warfare equipment for the Canadian forces. From heading a small academic department at McGill, Maass eventually found himself directing an office embracing a staff of two thousand men. In addition, he was chairman of the Advisory Committee of Industrial Chemists of the National Research Council, still head of the department of chemistry at McGill, director of the Pulp and Paper Research Institute, chairman of other diverse organizations, and member of still others such as the War Technical and Scientific Development Committee.

Medium in build and gentle of mien, Otto Maass did not find barking out orders the natural way to proceed, but he had one unique advantage: the corps of students he had personally helped to train to the doctoral level at McGill were now strategically placed in key chemical industries and university departments. When O.M. needed information, wanted something done, or needed any material, it could be arranged judiciously by picking up the telephone. Few people have greater influence than a former, well-respected teacher; the results were usually prompt and the best. 'This was a shock to military protocol,' Steacie later recalled, which required the correct administrative procedure at all times and strictly no 'going over anybody's head.'[10] To cap it all, Maass insisted on personally making hazardous trips to England, to find out at first hand just what was going on. In the same tradition as scientists such as J.B.S. Haldane, Maass exposed himself to mustard gas spray and other battlefield realities, rode in tanks during firing trials, and operated flame-throwers and the like. It all took courage. There were also numerous journeys to Washington. These personal contacts would initiate an unprecedented and perhaps unrepeatable era of co-operation, good will, and respect between American and Canadian military scientific efforts. So effective were Maass's methods around this time that apparently Germany, that most efficient of nations, hatched an elaborate plan to kidnap Maass and replace him with a substitute. Not surprisingly, and despite deservedly glowing reports of Maass's great endurance (typically a sixteen-hour working day, C.J. reported) in the face of seemingly impossible burdens, there was much sacrifice all around. His beloved chemistry department at McGill suffered, for one.

Maass's devoted labours on behalf of his country remain largely unrecorded; his passionate patriotism – which got people moving as much as anything else – was rewarded by public obscurity, the usual prize for those who prefer to work in the background, reticent in the face of acclaim. 'I never knew a more unselfish, courageous devoted Canadian subject,' C.J., himself a tireless worker, later commented to Steacie. Emotional defender of Canada's freedom to the end of his life, in 1961 Maass even took Lester Pearson, then leader of the opposition, to task for 'the Liberal party's tendency to go towards neutralism' – a fact earnestly refuted by Pearson in reply.[11]

But within the National Research Council Maass's efforts were greatly appreciated. Much of his success, according to C.J., could be attributed to his technique, which paid small attention to organizational or jurisdictional niceties but got on strenuously with the real job. By this means a number of the original time-consuming committees 'vanished in [an] apparently unrecorded and illegal way.'[12] Unorthodox maybe, but to those with too many missions to accomplish and too few hours with which to accomplish them, it was the only way. To Steacie, it was a lesson in people-orientation, observed with a keen eye and silent approval.

In the mean time, Mackenzie was establishing his own equally unique mode of government communication. The confusion of politicians responsible for what was what, who was who, and what was or was not to be designated a war asset, meant administrative road-blocks which absorbed time and energy. Still in his initiation period, sweeping general orders came through to C.J. which threatened to cut NRC expenditures to the 1936–37 level, just as programs were being projected for the critical 1940–41 period. 'Of course,' C.J. later explained, 'all such instructions were cancelled as soon as senior officials and ministers could be briefed.' C.J.'s and Steacie's personal abilities to put over NRC's point of view to these senior officials and ministers, and get results, would mark a special period of co-operation between science and government in Canada.

Mackenzie, emphatically non-military by nature, nevertheless laid his strategy for the battle ahead with much thought and tactical skill. To begin with, he acquainted himself thoroughly with the historical foundations of the National Research Council. Along with occasional excursions into the art of persuasion, Mackenzie now held the entire responsibility for NRC's diverse roles. He held the reins lightly, as was his style, conveying much of the time a distinct atmosphere of imperturbability. In part, this was achieved by delegating; in part, Mackenzie simply epitomized good sense, based on an innate understanding of human nature. The air of imperturbability no doubt owed much to his personal philosophy, learnt after his experience in the trenches of the First World War. He had decided then that whatever befell in the future, nothing in the way of misfortunes, violent confrontations, or frustrations would be permitted to upset him personally.[13]

Steacie, the younger by a dozen years, approached life differently. Problems were met by frontal attack, usually instantaneously. If initially this method did not meet with the desired results, he merely switched to an alternative position, and started the procedure again. The two men worked well together, perhaps because, while their personalities were markedly different, their purpose was intrinsically the same. In the end, their aims and achievements would be nothing less than autonomy for Canada in the scientific domain.

By 1940 all Europe and far beyond were swamped by war. Canadians, to their consternation, found that Britain, thus deeply engrossed, was inclined to dismiss Canada as 'no more than a useful subordinate in the scientific field, a country which might like to send over a few scientists and engineers to lend a hand in British laboratories.'[14] Actually, Mackenzie himself considered this a splendid course of action. Early in January of that year, he confided his thoughts to General McNaughton. 'There is one project on which I have been doing a bit of work, but on which it is difficult to get action. My suggestion is that arrangements should be made for Canada to keep in England an establishment of from twenty-five to fifty scientific workers who would be engaged in a civilian capacity in the various research establishments in England.' The plan would have two aspects. Certain workers could be placed in laboratories, where needed, for the duration of the war, where undoubtedly they could make a real contribution; others could be placed in various research institutions for a time and then brought back to Canada to assist in the work at home. The whole scheme seemed wise and sound. Canada would not only be making an immediate contribution of real merit, but when the war was over, there would return a number of bright young men who had gained valuable experience working under some of the best brains in England. The only problem with the scheme was that, as things stood, there was no channel by which NRC could carry it out. C.J. asked McNaughton to 'plant the idea in appropriate quarters.'[15]

In conjunction with this idea, C.J. was mulling over a more immediate and critical problem: how to mobilize for action the Canadian scientists and engineers who had offered their services to NRC as soon as war had been declared. The war itself, after all, was in Europe and communication was slow. The NRC knew nothing of strategic and tactical scientific needs for doing battle; it had been born after and only as a result of the last war, and times had changed. C.J. determined that the talents of these able people should not be wasted in work of uncertain validity or, worse, in work they would find, after strenuous efforts, was merely duplication. Part of C.J.'s problem, therefore, was to know where to start; what were the needs and how and if these could be met. As he explained, 'we did recognize ... that we [were] not a major power ... and we took a firm stand then that if we were to work efficiently, we must start on every problem with the full knowledge available to our Allies, and further that we must work co-operatively on common problems.'[16] Almost immediately, this principle proved impossible to sustain. As the war progressed, secrecy became paramount, stronger than all common sense which might have prevailed. It was a new experience for most scientists and one major drawback to war research; hardly anyone was supposed to know what anyone else was doing or why they were doing it. Understandably, this would throw a mighty spanner in many an operation.

Meanwhile, the imperial attitude of the British to the dominions was far from being unanimous. Hardly had the ink dried on C.J.'s communication to McNaughton when a pleasant letter arrived from Sir Edward Appleton, secretary of the Department of Scientific and Industrial Research. Sir Edward had heard of plans for co-ordination of research at the National Research Council and was offering his services. 'As your opposite number in this country, I may be allowed to say, without impertinence, that we here shall be glad to do anything that is appropriate at any time that you may wish to make contacts on this side ... I know that I can say this to you without having to explain that we have no wish to poach on your preserves or to offer gratuitous advice on the best way of doing your own job!'[17]

This was just the tone of letter that C.J. found helpful. In reply, he told Appleton that in fact he had been on the point of writing, and explained his philosophy. At the very least, the work done by the NRC 'should be appropriate and to the point.' No mention was made of the plan to second Canadian scientists to research centres in England, but C.J. had already discussed with Otto Maass the wisdom of having someone in England who would make personal contacts, search out what were the pressing problems, and 'investigate what contributions we in Canada can best make.' 'We will be only too glad to receive any suggestions that you care to make and you may take it that we will receive them in the spirit in which you make them,' he told Appleton.[18] Sir Edward Appleton, an affable, highly gifted scientist and Nobel laureate-to-be, had succeeded Henry Tizard as secretary of the DSIR in the same year as Mackenzie took over at the NRC. The two institutions shared a common purpose: the organization, development, and encouragement of national, scientific, and industrial research. DSIR, however, did not run laboratories of its own, although it did administer the National Physical Laboratory and was responsible for the geological survey of Great Britain.

By April the idea of having a permanent representative in London was firming up, though no final solution had evolved. Mackenzie wrote again on the problem of free communication between Canada and the old country to General McNaughton: 'we have tried many channels and while some of them appear to work we are never certain.' But fortuitously in May, Professor Archibald Vivian Hill, secretary of the Royal Society, Nobel laureate, and professor of physiology at London, now attached to the British Embassy in Washington, called on Mackenzie in Ottawa. Hill was quick to appreciate the problems and efforts at the NRC; by the time he left, he was 'full of admiration for what Canada was doing.' If only a man of Hill's calibre could be installed in Canada to act as a liaison officer, there would be 'an immediate channel to the scientific heads in the Old Country.' C.J.'s feelings on the matter, when discussed with Hill, were met with approval and the wheels set in motion immediately. Even so, it would be slow going.

In the late summer of 1940, Sir Henry Tizard's mission to exchange scientific and technical information with the United States took off from Britain. Tizard, however, chose to travel to Washington via Ottawa, no doubt after discussions with Hill. From then on events would begin to accelerate. Before the war, Britain had largely provided the advice, the men, and the model from which an institution such as the NRC had been created and the example which men like Macallum and Tory were inclined to follow. The emphasis would now shift to the new world, and Mackenzie and Tizard would be two major figures in this evolving scenario.

By the fall of 1940, Mackenzie was well accustomed to the pace which the war effort demanded. In July the corner-stone of the first NRC building (aerodynamics) to be constructed on the Montreal Road site was laid. Little could Tory have foreseen in 1932 that the huge building on Sussex Drive would not be able to contain the expansion of science that was to occur in less than a decade. Montreal Road runs east of the city and the new site, though separated by a distance of some six miles from the original building, allowed plenty of room for expansion. Ottawa in those days lay mainly west of the Rideau River. Today the growth of the urban area has spilled all around and beyond this eastern 'campus.' Already in Mackenzie's short tenure, the staff had grown from three hundred to five hundred. It would grow much more. Even so, organizing science for a war being fought across the ocean was difficult. Funds remained scarce and the Canadian government remained 'a bit confused about the scale of operation and with no intention of immediately mounting an all out war effort.'[19] A pleasant surprise materialized in 1940 to ease life somewhat and saved many a scientific project from dying before maturity – the appearance of a million-dollar 'Santa Claus' fund, the gift of a group of Canadian businessmen. By early 1943 the sum was $1.3 million – all to be expended on research projects under the auspices of the National Research Council, another triumph for C.J.'s powers of persuasion. To Steacie would fall the honour of finally dispersing this bounty.[20]

These were stressful times. The work of the Council increased minute by minute; days and weeks passed with no let-up in hectic activities. Strains began to show. S.P. Eagleson, secretary of the Council, was ordered away for a month because of illness; Otto Maass was on the verge of nervous collapse but Mackenzie could not get him away for a rest; E.A. Flood, the Council's expert on chemical warfare, broke down from overwork and had to take leave. C.J.'s natural buoyancy was sorely tried. 'Affairs here seem to move in cycles. At times we seem to be making excellent progress; at other times we seem to get bogged down by bureaucratic machinery and have a feeling that we are in a treadmill getting nowhere.' But as always, he ended on an upbeat note: 'Things are moving and I suppose our moments of depression are due to the fact that we all want to do so much more than seems possible to achieve.'[21]

But worse was to come. C.J. had made this comment in a letter he had dashed off to General McNaughton on 11 February 1941. The general was now in England and the letter was to be delivered personally by Sir Frederick Banting, leaving for England the next day by bomber plane. The famous discoverer of insulin and Nobel laureate was head of the NRC's Associate Committee on Medical Research, and a tower of strength to Mackenzie. But neither the letter nor Banting was destined to arrive in England. Banting died when his plane crashed in Newfoundland, and C.J. took the news very personally. 'You will appreciate more than anyone else,' he wrote to McNaughton, 'how much his loss means to the Research Council and particularly to me at the present time ... he has been ... a warm personal friend throughout the months.'

C.J.'s difficult task at NRC was greatly eased by his amicable relationship with members of the government and particularly C.D. Howe. When war had been declared, Howe had been the minister of transport, but by the spring of 1940 more urgent needs required the Howe touch. Obedient to his leader's wish, Howe took on the thankless task of minister of munitions and supply, without much enthusiasm. To begin with, there was the same question that had dogged C.J. in taking over a new department preparing for war – where to start. But once that was overcome, and it quickly was, Howe settled into high gear.

By the end of 1941, the NRC had increased to about one thousand employees, including those engaged in work across the country; even so, in September the following year, C.J. told McNaughton, 'we are arriving at the point now at the Research Council where we have more problems than we can undertake.' From time to time there were moments of triumph to alleviate matters. The distinguished scientist Irving Langmuir of General Electric come up to Ottawa in connection with his work on the problem of smoke penetration of gas masks. After a day spent with Flood, Langmuir had gone back much impressed with the work that was going on at the NRC. This, despite the fact that he had to modify his theories completely.

By this time the NRC had established warm relations with the United States, which after the attack on Pearl Harbor in December, was now deeply committed to the conflict. In fact, C.J. had formed a sustaining friendship and respect for the American working groups from the first meeting, during the Tizard visit of 1940. He had noted then that a National Defence Research Committee had been established, headed by Vannevar Bush, with an initial fund of some ten million dollars. Now, two years later, he observed that the 'three key men in the scientific hierarchy in Washington seem to be Dr. Bush, President Conant of Harvard, and Dr. Compton, President of MIT,' all of whom were 'most kind and generous to us in all our contacts.'[22] C.J.'s assessment of the three men was on target. In August 1943 the Quebec conference between Churchill, Roosevelt, and Mackenzie King ushered in a momentous era for the world. C.J. had been appointed the Canadian member of an international three-men technical committee formed to correlate the policy position of the three nations, now united by war. He frequently found himself with C.D. Howe in Washington, conferring with his American (Richard C. Tolman) and British (Sir James Chadwick) counterparts. 'We are on a most friendly basis with both the Americans and British and I think that I have been useful to the two parties in composing differences of outlook and opinion.' In fact, Mackenzie was understating his personal contributions.

By the beginning of 1943, C.J. and others were reflecting on the future of the NRC. Actually, the topic had never been far from Mackenzie's thoughts, and from time to time he had put his ideas on paper to McNaughton. Now that Hitler's might did not seem so threatening, C.J. felt a distinct lifting of the spirit. 'I do not know whether most other departments are experiencing the same thing, but the frantic period seems to be getting over in the Council.'[23] For the first time since assuming leadership, Mackenzie took a few moments to survey the scene from a distance; all appeared on track. Now would be a good time to plan a trip to England, 'to make personal contacts and see the various institutions in England in operation' – something, he confided to McNaughton, he had been wanting to do for the last two years. The trip materialized in May and June, and it confirmed and affirmed many a private conviction of Canada's relationship with Britain. Whatever it may have been before the war, things would be different afterwards. For a start, Mackenzie had the distinctly uncomfortable experience of telling Prime Minister Churchill that his dream of a great floating ice landing-field (code named 'Habbakuk') was impractical. 'I told him that there was no place readily accessible where one could get a seven or eight foot ice field and that you could not possibly freeze eight or ten feet of ice a day ...'[24]

Back in Ottawa, Mackenzie proposed that the next Review Committee in September be devoted to a brainstorming session for post-war plans. The direction of a national scientific laboratory like NRC was, of course, dependent on large numbers of economic and historical factors, many common to other countries. Something concrete to establish a few of these vital facts had already begun at NRC as early as 1942. The job of compiling this information fell to Frank E. Lathe, attached to Steacie's division. Lathe had preceded Steacie to the NRC by a generation, having been personally recruited by Tory some four years before the chemistry division came into existence in 1928. His subsequent tasks at NRC ranged widely, including the launching of the *Canadian Journal of Research* and acting as secretary to all important NRC associate committees throughout the 1930s, all of which did not allow much time for chemistry. It is interesting to note that one such as Lathe fulfilled these diverse and nebulous roles throughout the early years of the NRC's history: in an organization of some one thousand employees, administration as a department did not yet exist. For his role as NRC's factotum, Lathe had the distinction of earning a higher salary than Steacie, the division director.[25]

Lathe did his homework with care, and over a period of some months presented Council with reports on the development of the roles of science and government in major industrialized countries, including Germany, Britain, France, Japan, and the United States. Mackenzie no doubt made good use of these facts and figures to measure Canada's performance to date and its future outlook. When war was over he would need the back-up of these convictions to add to his powers of persuasion.

Steacie, too, had been compelled, for one reason or another, to give much thought to post-war matters. The chemistry division, itself a major player in the drama C.J. was co-ordinating, had since 1939 been buried in a plethora of practical problems, many of them to emerge as of considerable significance in the war. The staff was engaged in investigating methods of producing metallic magnesium and studying active charcoal; there were investigations into properties of textiles, explosives, petroleum, rubber, and refractories, and there existed a section on what was to become chemical warfare. By December 1941, 90 per cent of the division's work was devoted to the war effort, shared by a staff of about eighty, most of them temporary war appointments. Chemical warfare itself had passed over to National Defence under Otto Maass. Problems tackled by the division were diverse: paint and corrosion competed for attention with filter pads for gas masks, parachute fabrics with properties of fuse powders. And a major time-and-effort task was that of specifications for the departments of National Defence, and Munitions and Supply, to ensure Canadian production would coincide exactly with British specifications. There were investigations of laminated wood for aircraft construction, and on materials such as leather, plastic, rubber – physical chemistry now implied de-icing fluids and resolution of aerial photographs. In one year, Steacie told the Review Committee, the analytical laboratory had dealt with fifteen thousand tests on fifty-five hundred samples. Meanwhile, V.W.T. Scully, commissioner of customs, approached C.J. to take over the Customs Laboratory – since, as he told C.J., a small detached laboratory could not compete with an institution such as the NRC. Mackenzie asked Steacie to negotiate with Scully. To the chemistry division was thus added, in January 1943, the staff and laboratory of Customs and Excise with a budget of $100,000, dedicated to the 'examination of commercial products for customs and excise purposes.' In accepting this additional responsibility, C.J. had the future in mind, thoughts which he confided, as usual, to McNaughton: 'Many of the scientific departments of government, who are outside the orbit of the Research Council, [feel] that they would like to be within our organization. I feel that after the war there will probably be some comprehensive reorganization of all departmental scientific activities here.'[26]

And what of the 10 per cent of the chemistry effort not devoted to the practicalities of war? This was largely accounted for by Steacie's own little unit investigating photosensitized reactions and Leo Marion's work on alkaloids. Despite all that was happening, Steacie managed to salvage some time to carry on his search for an understanding of chemical reactions. On train journeys, between tasks, and especially at night,

Steacie studied the scientific literature, puzzled over new data, formulated explanations, and wrote up the results. Twenty-three research papers and a 250-page book would come from these snatched periods of industrious labour. It was a particularly interesting time in the development of reaction mechanisms; if neglected, no scientist could hope to make up for lost time. This was true not only in his subject, but in science as a whole, and Steacie undoubtedly wanted a role in it.

During C.J.'s absence in England, Steacie and Robert Newton, a member of Council, had put their heads together to respond to communications from the minister of trade and commerce and the Canadian Chamber of Commerce on the question of new products likely to affect post-war trade. Steacie had put synthetic rubber and fibres (including glass) at the head of a comprehensive list which would have made any industry-minded minister happy.[27] Before the next Review Committee meeting, scheduled for 16 and 17 September, Lathe sent Steacie his own memo regarding post-war research, suggesting a number of areas, such as pure alumina, slag studies, and wool growing and manufacture. Lathe also came up with the suggestion that research in Canada would benefit from decentralization. 'Many problems can best be undertaken locally, but require for their solution the advice and assistance of experts from a central organization in which fundamental research in all the major branches of science is carried on and where the most complete library facilities are available.'[28] This part of the memo struck a chord with Steacie, for he put it forward as the first item on the agenda when the Review Committee convened. Although the first to address the meeting, Steacie intended only to give an overview of chemistry at the NRC; as for the actual areas of work done within the division's auspices, he wanted the men heading the sections to speak for themselves. Steacie believed in delegating and placing total confidence in his chosen men; it would become a hallmark of his leadership.

First, then, Steacie thought the work of the laboratory needed to be decentralized throughout Canada, so that specific industrial problems could be dealt with on the spot. In the past few years the division had become almost, though not exclusively, a source of consultation and a testing laboratory for both the armed services and industry. These questions and interruptions, if continued beyond the war period, would disrupt any in-depth research work being done. In the post-war period, Steacie advocated that Council should strengthen scientific research in universities, support industrial research, and see that a fair proportion of basic work was carried on in the chemistry division.

It was on this occasion that Steacie made his first policy statement, which would have long-term repercussions on Council laboratories. In his opinion, it was desirable that 'the staff engaged in these two [basic and applied research] fields should be separated.' Fundamental research, he thought, would largely be the domain of workers at universities, and industrial research would be maintained at NRC. To date, the problem with research in industry was that Canadian companies were largely branch plants and merely used the results provided by their parent firms. It was important, Steacie maintained, to provide opportunities for scientific research at the NRC, if the Council was to attract competent workers to its staff. As an example of pure research which should be pursued, Steacie suggested light metals and alloys. But above all, Steacie was for clear thinking about science and its direction. As for the kind of questions just put to him by a parliamentary committee, on whether a chemical industry could be established at a location on the basis of its coal, Steacie could not refrain from showing his contempt for such a piece of illogicality; to his mind, that was just another example of political expediency. Coal was hardly an important factor in most chemical industries. This type of talk, if taken further, would suggest that 'as there were large quantities of building stone in Ungava, buildings should be erected there to utilize the stone.'[29]

Steacie's stand on these issues marks indelibly a watershed in NRC development. Although his views did not focus until two years later, the die was cast. In time the chemistry division was divided, as he had suggested, and Council's support for university research increased dramatically. More important, Steacie had opened a Canadian debate that would persist and reach a crescendo after his death: the question of NRC's role in so-called pure versus applied research endeavours. The same ball would bounce back and forth into different courts, with no ultimate resting-place.

Mackenzie was also in the process of working out his thoughts on the whole matter of reorganization. As leader of the institution, he was faced with broader facets of the future. 'The picture in brief,' he told McNaughton in September 1943, 'is that the work of the Research Council during the war years has created a favourable impression and the government as a whole is kindly disposed to us and rather favours all scientific work being done under our auspices.'[30] This plan, if implemented, would require considerable reorganization, and above all, continuation of government good will. That year, the political situation did not appear at all certain; the government, after losing four by-elections in a row, was understandably perturbed and rumours abounded on the possibility of a general election. All of which gave Mackenzie much cause for concern: 'If an immediate election were called, the chances are that the CCF might be returned, if not with a majority at least as the largest group. When one considers the chaos that would exist while inexperienced people were finding their feet, one cannot help toying with the idea that it might be better to get our future settled while a government is in power that understands generally our needs.'[31] Mackenzie need not have worried. The Liberal star was still rising, and for the next decade and a half at least, government relationship with the NRC would be uniquely sympathetic. Good days were still ahead.

There was one other related task for Mackenzie to perform before the end of the year. In October Howe asked him to prepare an outline of the plans and discussions held at the NRC. This was to contain major policy directions of national importance, so much so that when Howe (once more reluctantly) accepted the post of minister of reconstruction exactly a year later he 'insisted [Prime Minister] King add the National Research Council to his responsibilities.'[32]

With eyes and mind set so much on post-war reconstruction, it might easily be forgotten that the real issue of the day was still war. The weeks and months of battles in Europe followed one another, though not fast enough for most. There were moments, too, when Steacie wondered what he had let himself in for: did laundry committees and refractory properties alone – important though they were – spell the way to the future of chemistry at NRC? What of the role of scientific research which had now been largely shelved, the understanding of scientific problems as yet unresolved? Was Canada and the NRC to have a role in this development? If so, how and when? Difficult questions to answer, yet too critical to ignore. But for the present, more urgent events demanded everyone's undivided attention and full energy. Steacie pushed these thoughts to the back of his mind and did not allow them to surface again until after the business of war had been dealt with. It would be seven years from the time when he joined the NRC before he was finally able to return to his role purely as director of the chemistry division in Ottawa. By then his first term of appointment at NRC was over and it was time to seriously consider a second, or to return to teaching and academia. In the mean time, the NRC had entered a totally unforeseen aspect of war in which Steacie would play an unexpected role. Early in 1944, a chemical engineering section was created in the chemistry division. This would become the hub of heat-transfer pilot plant studies for the nuclear pile at Chalk River, providing both staff and facilities for an NRC venture to which we now turn.

6

Lessons in Scientific Diplomacy

Of all the diverse aspects of chemistry embraced by the division during the war years, nothing was more unexpected than its involvement in a field hardly in existence at the time, in a project which would have revolutionary consequences. On becoming director, Steacie became entangled in nuclear and atomic issues almost immediately, a continuation of events already in progress under G.S. Whitby. It all began innocuously. Canada's happy position in possessing a rich source of pitchblende meant that the much desired element, radium, was readily available. After Madame Curie's famous 1898 discovery the uses for this scarce curiosity appeared endless. The chemistry division was, by 1939, routinely testing products such as the bromide of radium extracted from the pitchblende by the Eldorado Gold Mines Limited.[1]

A vast tonnage of the ore, however, produced only minute amounts of the precious radium, as its discoverer learnt to her cost. By far the major products from the ore were compounds of the element uranium, isolated in 1841, for which no great use had so far been found. Inevitably, vast stockpiles of this material were creating a nuisance for the mining operation, though some sort of outlet had been found by using it for tinting glass and ceramics – hardly a drain on the mountainous waste pile. Even the promised increase in demand for uranium oxide for souvenirs marking the coronation of George VI did not help. Suggestions as to its potential and wider utility were quite ingenious: could a uranate salt be used as fertilizer? what about the rubber industry, to reduce or even prevent static electricity? perhaps tinting the glass of traffic lights?[2] Conspicuously absent in all these ponderings was the notion of radioactivity.

These questions were now all in Steacie's domain. D.F. Stedman, an inveterate innovator in the chemistry division, had come up with the idea of making uranium alloys, which might perhaps be useful for the steel and nickel industry. In the late spring of 1940 Steacie cabled Otto Maass, who was in England studying the possibilities of co-ordinating Canadian and British scientific programs. The previous summer the Admiralty had indicated an interest in uranium steels to General McNaughton, and Stedman had now successfully produced good ferro-uranium by an economically viable method. Were they still interested, Steacie asked. 'If not, can they give us suggestions regarding future work?'[3]

It was not to turn out so simply. Unbeknown to Steacie and Stedman, scientists elsewhere were already turning toward a totally different idea for the use of this by-product. Almost simultaneous with Steacie's cable the press began buzzing with reports that United States and Nazi scientists were racing to get control of a new superpower substance, 'one pound of which is said to be capable of yielding power output of 5 million pounds of coal or 3 million gallons of gasoline.' This miraculous substance, they declared, was the nearest approach yet to that eternal hope of humanity – perpetual motion. Was the solution finally at hand? 'As long as the Uranium-235 is supplied with water it would keep on liberating its energy until it is exhausted.'[4] The excitement had special significance for Canada. Mines and resources reported that if uranium 235 were developed sufficiently for commercial and military use, it would be a boon since Canada was supposed to possess most of the world's uranium ore. As one of the most knowledgeable on the subject in Canada, physicist George Laurence remained sceptical, particularly to the suggestion that the release of the purported energy could be achieved 'even with explosive violence.' The possibility was not ruled out but it seemed dubious.[5] Yet it was curious how many people were becoming interested in this substance. Even an experienced scientist such as John Cockcroft was writing personally in the same month to Boyle of the physics division, asking what stock of uranium oxide ore existed in Canada and whether he knew if the United States was already producing the metal.[6] The reason for these questions was revealed only much later.[7] The National Defence Department also began to show interest. Could NRC advise them on the potential uses of this new commodity? It was not an easy question to answer. Scientists were stuck with the problem that no simple method for separating the two naturally occurring uranium 235 and uranium 238 isotopes was yet known and only the first substance was of immediate interest. The difficulties of achieving this separation would be formidable, but that did not prevent ingenious individuals from having a go and reporting success.[8]

Maass's reply to Steacie's cable was perplexing: 'Very secret,' it ran; 'how many pounds of pure uranium could be sent to United Kingdom for experimental purpose if required?' Steacie pondered the significance of the request for pure uranium. The material, Steacie cabled back, was in the form of 30 per cent ferro-uranium; present production was laboratory scale and uranium was expensive. Larger production had to depend entirely on British demand. By 7 June, Maass had still not received Steacie's cable and sent an urgent message to hasten a reply.[9] Events now moved rapidly and attitudes changed sharply almost overnight. Earlier, the high commissioner in London, Vincent Massey, had replied calmly to Canadian inquiries regarding the possibility of British interest in this uranium steel: the armed forces could not express any decided views until more was known of the properties, cost, and availability. One hundred per cent uranium, however, 'had some interest.'[10] Six days later, the director of scientific research at the Ministry of Supply asked Massey to cable urgently, 'to obtain for us full details of Canadian technique for preparation of pure uranium metal.'[11] Steacie decided that full details could most effectively be sent by letter, but a quick cable should emphasize that the process was for ferro-uranium, not pure uranium. In the end, a brief account of Stedman's method was sent with a cautionary note: pure uranium was very hard, forming objects from massive metal would be difficult, 'if you wish us to take matter up cable further details of uses of metal or form in which it could be used and amount which might be needed.'[12] Even now, Steacie did not suspect the impossibility of this suggestion, but a hint of the direction in which things were moving came just a short time later.

By July, Otto Maass was back in Ottawa. Together with G.H. Henderson who had accompanied him, Maass wrote to C.J. On his recent mission to England, he and Henderson had learnt of the potential use of uranium as a military weapon. The demand for uranium both from the United States and from enemy countries was now heavy. He recommended that export of uranium from Canada should be prohibited until further clarification.[13]

If the group in Ottawa were now fully aware that scientists and the military alike had a special interest in uranium, events in Europe had been moving much faster. In the late summer of 1940, John Cockcroft arrived in Ottawa after accompanying Tizard to Washington on the first U.K./U.S. scientific exchange mission. Cockcroft, after a brief hesitation, showed a knowledgeable and friendly concern for Canadian work in the nuclear field.[14]

The war in Europe and beyond worsened. It was rapidly becoming clear that this was a new kind of war, with huge implications for science, and it would require the co-operation of the political, military, and scientific leaders of many nations.

Scientific co-operation in this new field began promisingly enough. By late 1942 the NRC's involvement in nuclear research was well and truly launched, but it was not to be smooth sailing. As things had stood a year before, Britain was already advanced in both uranium research and the first stages of it use.[15] Henry Tizard's ability to confide the British state of affairs to American scientists and decision-makers Vannevar Bush and J.B. Conant had brought positive responses; it was a time of understanding between diverse statesmen of science. They had concurred then that a joint Anglo-American effort would bring faster results. The mixed team of scientists now working at the Cavendish Laboratory in Cambridge on problems of a heavy water reactor could advantageously be relocated in the United States. But by early September 1942, it had been decided that Canada would be their new home. On 24 September the leader of the team, Hans van Halban, arrived in Ottawa to call on C.J. Mackenzie.

Halban's team would consist of three groups, one each in chemistry, physics, and engineering. There was to be a technical direction committee, consisting of one from each of the three groups with Halban, George Laurence, and one other, to be nominated by C.J. Mackenzie already saw some of the problems ahead: 'I stressed very strongly that we could only have one single authority in matters of policy.'[16] From the American vantage point things were beginning to look less congenial than a year earlier. Although Halban may have wished the program to be controlled only by Britain and Canada while drawing on the resources of the United States, it is certain that the international team of scientists being gathered in Canada did not see things this way. Nor, indeed, did C.J., who knew better than anyone what the British and Americans did not know about each other's work. The Americans would not be likely to provide heavy water, still less their knowledge. The technical problems of building a reactor, great though they were, would be as nothing compared with the political problems.

For such a new enterprise, where to locate the nuclear laboratory was quickly decided: it was to be Montreal. Halban's group set about recruiting scientists for the difficult job ahead. French exiles Jules Guéron, Pierre Auger, and Bertrand Goldschmidt were among the earliest to take up residence in Montreal. Even the inconvenience of being crammed into an old house, where space was strictly limited and no laboratories existed, failed to detract from the air of high expectation. Something new and phenomenal was taking place in science and everyone on the team was out to play a part; for some of the younger members, it would be the most exciting period of their scientific careers. When the move was made early in 1943 to a brand new wing of the University of Montreal where space became 'semi-infinite,' the excitement grew greater. It was a fine laboratory, standing on the north side of Mount Royal. Each day, somebody or something different arrived to join the activities, adding to the fresh yet urgent beginning. Equipment was modern and the best available; when an over-zealous English technician mistakenly ordered 24-foot bellows for glass blowing, the marvellous efficiency of the new methods and the redundancy of the old seemed complete.[17]

On the nuclear chemistry front, things were moving particularly fast. The 1940s had begun with new discoveries, ingenious techniques of analysis and identification, and more challenges than seemed possible to meet. By 1941 a team at Berkeley (including chemist Glenn Seaborg) had proposed the theory that the 239 isotope of plutonium, discovered a year earlier, was capable of sustaining an explosive chain reaction. Early the following year the Americans decided to concentrate on developing methods to isolate the plutonium isotopes after scaled-up production. The chemistry team from Berkeley would join physicists Arthur Compton and Enrico Fermi in Chicago. It was early days yet in the game and scientists were still relatively free. On 17 April 1942, Seaborg boarded the train east carrying the world's total supply of plutonium in his briefcase.[18]

Before arriving in Montreal, Goldschmidt had worked on radiochemical problems in Chicago, where his presence had been warmly welcomed by Seaborg's group.[19] The main business there had been to learn how to extract plutonium efficiently from the spent uranium reactor fuel. Goldschmidt had helped to map the large number of isotopes, the products of uranium fission. He then turned to more chemical endeavours, helping with the preparation of the first sample of a pure plutonium compound, a triumph achieved in August 1942. The Montreal lab thus had in Goldschmidt an experienced and gifted nuclear chemist, one of those rare and intuitive souls who could 'feel and taste chemical properties.' The chemistry group, headed by veteran chemist F.A. Paneth, did not lack for talent; there simply was not enough to satisfy the task at hand.

The genuine excitement and abilities of the assembled group notwithstanding, trials and tribulations materialized very quickly. Halban found the organization of the team, small as it still was, difficult from the beginning. He lacked administrative experience and proved particularly inept at dealing with bureaucracy which was, after all, a 'permanent part of government organizations,' as C.J. told him. There was no getting away from Treasury Board restrictions or any other of the petty annoyances. That Halban was finding it a trying time is understandable. Dealing with the realities of a top secret scientific enterprise where the final word was not only not in his hands, but not even in Canadian hands, was extremely frustrating. Even C.J., that most imperturbable of leaders, was beginning to find the situation irksome. 'The trouble with the Montreal group is that their minds work too fast. They are always in a stew about something and they are willing to try anything without thinking about it and thinking it through.'[20]

Mackenzie meanwhile had plenty to give him thought. A letter from Conant spelt problems of the worst kind for the Montreal group. Scientific confidences and exchanges between them and their American counterparts, as earlier planned, would not be permitted. After counselling Halban not to act too hastily, C.J. departed for New York, to return to Montreal just three days later with encouraging news. Conant, perhaps more aware than most of the distressing effect of his message on the Montreal team, had asked C.J. if Halban could meet with Harold Urey and Fermi in about two weeks' time to discuss their research. Scientific hearts lifted perceptibly.

Conant's letter to Mackenzie had come as a particular blow to the chemistry group in Montreal. Physicists were still free to exchange information with Compton's group in Chicago, as far as the use of heavy water was concerned, but 'not in regard to the chemistry of element "49" or the separation.'[21] However the picture may have appeared to the powers in London and Washington, the entire group in Montreal now saw their past efforts and future work tottering on the brink of dissolution. The recent high expectations contrasted sharply with their present feeling of defeatism, tinged with defiance. The scientists felt they had to carry on, on their own if necessary. Affairs, though, had not deteriorated so far as to negate personal contacts at the scientific level. So both Auger and Goldschmidt decided to risk a meeting with former friends and colleagues in Chicago.

Their reception by the Chicago group, also labouring under difficult conditions of secrecy, was warm. Auger and Goldschmidt returned to Montreal that same evening bearing precious gifts: experimental data and two tubes containing a small quantity of vital fission products and about four micrograms of the plutonium Goldschmidt had earlier helped to extract. 'The line had been crossed,' Goldschmidt was to recall, 'my scientific career had led me to commit a political act; already the atomic race had begun – between allies.'[22] In retrospect, the determination of the Canadian team to go it alone in developing a process for the extraction of plutonium – with the help of the contents of Goldschmidt's precious tubes – proved a blessing in disguise. Self-reliance meant doing their own thinking on how to separate plutonium and uranium and would lead to successful new methods of solvent extraction for the highly radioactive fission product, by remote control.[23]

At the human level, however, things had remained cloudy. That Halban knew little about administration was now painfully clear and his communication with the chemical staff – or lack of it – was something less than ideal. When informed, C.J. had been sympathetic and took Paneth to lunch to discuss the problem; Paneth was coping as best he could but apart from anything else, the lack of chemists on the team was becoming critical. C.J. suggested using the talents of young men such as Harry Thode and John Spinks.

Things continued to move slowly. This time the solution seemed obvious to Mackenzie. 'I suggested to him that he should have Dr. Steacie associated with his work and he was more than anxious to do this and has already made arrangements with Dr. Steacie.'[24] Of course, the chemistry division in Ottawa already had a full load and as director Steacie could not drop these responsibilities. His role in Montreal would be in some part-time capacity. Certainly he was in a better position than most to advise Paneth on acquiring Canadian workers, but C.J. stressed 'if Dr. Steacie becomes associated with the project he has made it perfectly clear that he does not wish any executive or administrative authority whatsoever and will consider himself as working entirely under Dr. Paneth's supervision and direction.' The team in Montreal appeared to welcome the good news, relieved that there was now a senior Canadian chemist associated with the project.[25]

So it was that Steacie, in early 1943, entered the lives of those working on atomic and nuclear problems in Montreal. Once launched onto the Montreal scene, it became quickly evident that things would not work out quite so simply as he had hoped. There was positively no getting way from the grind of his executive and administrative responsibilities. As Mackenzie's direct appointee and director of NRC's chemistry division, Steacie had powers in matters over which Paneth, essentially a British-employed scientist, had no jurisdiction. Apart from the shortage of chemistry personnel, there were the problems of those already there. Cost of living in Montreal was rising, and for some of the younger members the scale of salary was becoming a real issue. Steacie took stock of the situation. In his forthright manner, he told those affected that he would see their salaries would be increased and by how much. More remarkable still, the promise was implemented without delay. To those whose lives had been something less than contented and harmonious under the heavy hand of Halban, whose own problems were multiplying rapidly, this calm, direct approach seemed nothing short of miraculous. After weeks of uncertainty, instant action and immediate results; once more there was a perceptive lifting of the spirit, especially among the younger corps. Having Steacie to turn to was 'like a breath of fresh air.'[26]

The problems to be resolved were not all so straightforward as that of salary, but Steacie viewed the turmoil pragmatically. Some difficulties were not unlike those back at his own quarters in Ottawa or, for that matter, the problems of academic administration he had faced at McGill. Some aspects of the situation, however, were quite novel. The experience of soothing the ruffled lives of an international bunch of prima donnas was not easily forgotten by Steacie, but he kept no record of his trips to Montreal nor details of developments there. However, his reluctance to take on any more executive authority was not in any way coloured by the slowly simmering milieu of the Montreal operation. That did not appear to disturb him unduly, and no one in Ottawa remembers Steacie being particularly perturbed before or after one of these visits. They were proving useful and he accepted them as an addition to the war effort.

The pattern now for Steacie was to go up to Montreal a couple of times a week or whenever necessary, sometimes staying a few nights at a time. This routine had its compensations. It gave him time to think about the problems of chemical kinetics; this was where his real interests lay, but the intensely busy days now left him little time to ponder over them. So he was able to put the journey between Montreal and Ottawa, up to three precious hours each way, to good use. As for the beleaguered group of scientists he administered to, Steacie's ability to assess diverse complaints and transform them into a series of priorities became something of a phenomenon. It would put him into many a young worker's personal 'hall of fame.'

For Steacie, by far the most enduring lessons of this episode had nothing whatever to do with chemistry, atomic or nuclear. In years to come, strong feelings and even stronger opinions were to emerge on apparently unconnected issues – precisely the issues which now bubbled and boiled in the Montreal setting. As Steacie made his reports back to his leader at NRC, he became increasingly aware of the Anglo-American political tangle and of Canada's difficult role as mediator. He observed the destructive nature of secrecy enveloping these scientific efforts and its effect on the morale of workers. Halban's personal insecurity merely multiplied the problems of the laboratory, adding to the already unhappy atmosphere brought on by strict regulations. Less senior scientists in different groups were forbidden to share their knowledge; code words and names abounded. Circulation of reports was restricted by the simple procedure of stamping them 'secret' with the added admonishment L (Limited) or LL or even LLL. The Montreal team began, as Goldschmidt put it, 'to be less anxious to know about the results of scientific experiments than about the results of negotiations between British and American leaders.'[27] For it was these political decisions that affected whether the very work they strove to accomplish continued or came to a full and premature stop, yet they were decisions over which, as scientists, they had no jurisdiction.

Steacie observed all this in passing and over the years the experience would linger and develop into a lifelong abhorrence for any aspect of science not open to discussion or over which scientists themselves had no control. He also saw with great clarity the need to remove the stresses of day-to-day organization from the shoulders of those labouring to do important and creative science. This, above all, was to become the focal point of his administration a decade later. To those around him – to C.J. in particular and perhaps even to himself – Steacie demonstrated a natural flair for the dreaded administration which invariably accompanies the organizing of large-scale science. True, when he had taken over the division directorship he had very quickly made a number of changes and set about reorganizing his domain, but with the announcement of war almost simultaneously, his own plans had had to be shelved. The job in hand – bringing order out of chaos in this unexpected chapter of NRC work – was categorically unrelated to his idea of doing science. His predilections for science did not stretch to pushing memos, but if dealing with people and their problems was the necessary stuff on which laboratory efficiency depended, he would do it. Relatively speaking, it bothered and taxed him little.

The Quebec Conference in August 1943, which Mackenzie attended, finally brought news. Scientific co-operation, of sorts, between the United States and Britain would resume and construction of the heavy water reactor on which the Montreal laboratory had been working would go ahead. All sides agreed on the appointment of physicist John Cockcroft as director of the new enterprise. Steacie formally accepted the role of assistant to Cockcroft; Halban became a figure of the past.[28]

Cockcroft arrived in Montreal in April 1944, to pouring rain. The dampness of the occasion by no means affected subsequent events; in the laboratories the mood took an upward swing. For one thing, Cockcroft at once set to work dismantling Halban's over-zealous concern for secrecy. Reports which had previously borne the officious Ls were given an airing. Steacie found in Cockcroft a kindred spirit in his manner of doing things, which meant brief and to the point. This attribute helped expedite matters greatly in the days to come, for there was much to do.

A major part of Cockcroft's efforts centred around negotiations. Even now, there was reluctance to share information on the chemical separation of plutonium although the Americans did agree to provide irradiated uranium slugs so that the Montreal team could make their own way in the separation process. A few days after arrival in Montreal, Cockcroft was off to see Compton in Chicago, Chadwick in Washington, and finally to Oak Ridge where the first graphite-moderated reactor was now fully operational. Delicate negotiations also necessitated trips back to England. When Cockcroft was away and matters required, Steacie took the lead. By winter of that year he found himself more deeply involved than ever in the affairs of the Montreal lab and the newly proposed reactor. He took charge of the new group starting work on reactor problems and was also expected to 'coordinate all the work carried out in the experimental water treatment lab, including heat transfer and corrosion.'[29]

Meanwhile, late in May Steacie joined Cockcroft in the discussions on sites and plant requirements for the proposed reactor. The location was critical: it required isolation in case of accidents but also the facilities of a town nearby; plentiful water for cooling and to carry away effluents; several thousand kilowatts of electricity; and a supply of manpower and therefore good transport and reasonable access to centres in Ottawa, Montreal, and Toronto. Even these conditions did not exhaust the list. The search ranged far. Nobel on Georgian Bay was 'delightful'; the Mont Tremblant region was equally delightful and Cockcroft relished the experience there of 'blazing pine log fires and good food'; the St Maurice River area provided 'idyllic surroundings' in the midst of fishing and hunting country.[30] But none of these turned out to be completely eligible. Later in the year Cockcroft and Steacie travelled to Pembroke and twenty miles beyond to Chalk River and Balmer Bay on the banks of the Ottawa River. This time Cockcroft was truly delighted: 'We collected wild strawberries and fell in love with the place, there was ample water and plenty of space and good communications. We quickly picked on a site a few miles downstream for the plant and a very delectable site a few miles upstream for our village.' That accomplished, the next step was to convince C.D. Howe; initially reluctant, Howe gave his approval.[31]

Work on chemistry, physics, and engineering ploughed on in the Montreal laboratory, where Steacie was now spending most of his time. Halban's departure in 1944 had the advantage of bringing his estranged colleague Lew Kowarski from Cambridge. He now became leader of the group, which started on building a small experimental pile, in anticipation of the more ambitious NRX reactor. Named ZEEP by Kowarski, it was completed in just over a year.[32] With the departure of several more of the European scientists, Canadians began to make up almost half the lab staff. When Paneth also returned to England, Steacie took over as head of the chemistry division at Montreal. 'We have now collected a good staff amounting to about 130 graduates,' Cockcroft wrote to his brother, 'we have seven New Zealanders, four French, 22 Cambridge [and] six Oxford graduates among the UK contingent.'[33] Growing involvement now necessitated the setting up of an extramural division, to oversee all the work done at NRC in Ottawa for the Montreal project in industrial chemistry and chemical engineering. Steacie was also made the director of this new division. By the spring of 1945, his preoccupation with Montreal had become so dominant that the chemistry division in Ottawa was having to cope without his presence much of the time, and had had to appoint an acting director, Adrien Cambron. From Montreal Steacie organized the 'loan of men' from industries and searched for other needed manpower.[34] His timetable began to resemble that of an administrator more and more. In the second week of March 1945 he was in Montreal; the next day he was back in Ottawa; six days later he would be at the Chalk River site to discuss the properties of lining for concrete tanks.[35] Construction of the plant at Chalk River was progressing, but with the usual hold-ups. 'I have got an operating date of the 1st of May 1945 – one that will need hard going to meet,' Cockcroft told Edward Appleton. 'Canadian industry is now our weakest link.'[36]

Cockcroft, with characteristic enthusiasm, threw himself whole-heartedly and equally into the nuclear project and the unique experience of a Canadian winter. It was a severe one and snow was plentiful. The Cockcrofts had the pleasure of learning to control skis, encouraged no doubt by Steacie, a keen skier and jumper. The frequent trips to United States continued and in April 1945, Cockcroft departed again for

England to discuss Britain's ongoing work in atomic energy. For the Montreal team and NRC as a whole, the post-war phase was fast becoming a reality. C.J. had already held discussions with Cockcroft and Steacie the year before on this issue, and that December there had been more discussions on the future. 'Steacie and Cockcroft think we should build up laboratories in Ottawa when Montreal closes down,' C.J. wrote, 'Laurence wants everything at Chalk River.'[37] But when the news of Cockcroft's impending return to England for good came, Mackenzie was understandably perturbed. Such a step seemed both unwise and unfair. The Chalk River reactor was still in its infancy. To lose its director would bode ill for the health of the program. C.J. took the unprecedented step of offering Cockcroft the as yet non-existent post of vice-president of the NRC, a singular honour and an enticing one. Cockcroft was genuinely tempted; he was enjoying the Canadian experience, but England, he was urged, had need of him.[38] In September 1946, Cockcroft's role as director of the Montreal laboratory came to an end and another Cambridge man, W.B. Lewis, took over at Chalk River.

By the fall the whole reactor project had expanded to some ten thousand acres and employed some four hundred scientists and engineers, a figure outnumbered only by the support staff. Things were running smoothly enough, and Steacie felt that the time had come to give up the extramural responsibilities he had acquired since 1943. He sent a memo to C.J. Mackenzie. 'In view of the fact that Dr. Lewis has now taken over the direction of Chalk River, that the project is now a purely Canadian one, and that I have returned to my former duties in Ottawa,' he wrote, 'my position as Deputy Director of the Atomic Energy Division is an anomalous one ... I do not feel that I should occupy a position which appears to give me administrative responsibilities which I neither want, not in reality exercise.' Still, if called upon, he would he happy to co-operate with Dr. Lewis in every way and continue to act as a consultant on chemical affairs.[39]

In all, Steacie's formal association with atomic and nuclear problems had covered just over three years. He had entered with a serious plea not to be placed in a desk-bound, paper-laden position and departed still maintaining his general abhorrence of administration. Yet his role in the whole Canadian undertaking – unrecorded and apparently not wholly appreciated, even by some of the figures involved – had not been without significance. Out of this interlude came his first real experience of administration within a large and complex scientific organization. The period had also provided his first glimpse into the way science, government, and politics interacted, or failed to do so. The experience had given him moments of both joy and tribulation, and in addition, there was the sheer novelty of this new chapter of science and Canada's role in it. The whole experience had convinced him of Mackenzie's great ability as the NRC's leader. To be launched into this tremendous war effort together had been a unique beginning to a friendship that would grow in the future.

Mackenzie maintaned a keen eye for appraisal at all times. Although Steacie's innate abilities were now quite evident, there were some edges to be smoothed out first. Besides, the NRC was blessed with a number of gifted personalities. When Robert Newton, director of applied biology, left the Council laboratories for Alberta in 1941, Mackenzie had debated the question of his successor with McNaughton. 'After considering the matter carefully and observing different people, I feel that Dr. W.H. Cook is probably the ablest man whom it would be possible to get for this position,' C.J. mused. 'It is true that he is comparatively young but he is as mature as Dr. Steacie, if not more so, ...'[40] By the end of the war Steacie had no doubt matured greatly.

All these developments were, of course, eclipsed by the momentous event which shattered the scene in the late summer of 1945 and forced the world to pause. When it recovered it was to a new era.

Mackenzie and a few of his staff had known of the impending use of the nuclear bomb, and six weeks before the event were already quite prepared. On 13 July C.J. had met with senior colleagues, including Steacie and Laurence, to discuss a suitable press release. Aided by Cockcroft, a short final draft was now ready. C.J. retreated to the peace of the Laurentian hills – his first real rest since the war began – and waited for the rush of the press and barrage of questions which would surely follow.[41] On 6 August the bomb exploded over Hiroshima.

Much to C.J.'s surprise, the press did not appear much excited. The secrecy surrounding the whole atomic enterprise was obviously having the desired effect; the significance had yet to sink in. No doubt, too, the impact was diminished by the excitement of a Japanese surrender and the realization that the war was over. The press finally awoke when C.D. Howe called a press conference a week later. Back at the Montreal laboratory, the reception to the news of the Hiroshima and Nagasaki bombings was more measured and sombre, stemming from greater insight into the significance of what scientists had done. Many had known, some had suspected, that the bomb was near completion in the United States, but the debate until then had centred on whether it would work. The actual event created both elation and depression for the Montreal team, not unlike the experiences of their counterparts in the United States.[42] Of course, the Canadian team had not taken part in the actual construction of the bomb; but for many, even apart from those who had conceived the original idea of fission and brought a scientific theory to fruition, life would never be quite the same again. Scientists were at the heart of this new creation but few, if any, were to have a say in its fate.

Science and politics would never again be entirely separable or at ease together. At the beginning of the work there had been few conflicts – only a feeling of urgency, a large dose of scientific curiosity and excitement, and even a keen sense of competition. The evils of aggression which had threatened the world were paramount and had to be contained. Scientists themselves had hardly had time to understand the real implications, though many had doubts. 'The only hope,' Cockcroft explained in a letter to his mother, ' of bringing politicians to their senses was that provided by the new power.' Cockcroft's view was shared, doubtlessly, by the majority of scientists then, whether they worked in the atomic and nuclear field or not. But the corollary to this thinking, also largely shared, was to prove a fundamental departure from reality, as it eventually evolved. 'I really do believe,' Cockcroft wrote, ' that we now have the choice between the "big three" living peacefully together, or of an annihilating conflict.' This feeling of confident superiority extended equally to those holding political power. The atomic bomb, President Truman decided on hearing of its successful birth, was too dangerous to be let loose in a lawless world. 'We must constitute ourselves trustees of this new force – to prevent its misuse and to turn it into the channels of service to mankind.'[43] The problem was, of course, that this lawless world did not agree as to who should, or should not, be the trustees of this power.

For the time being Steacie was not personally affected by these world events, but their consequences would in the near future become a major part of his domain. What had begun as a scientific enterprise was now a political tool. Soon it would also provide a leverage for science. Steacie had made his contributions to Canada's nuclear program without protest or fanfare. He had been as efficient and as helpful as he was able. No doubt, he was in total agreement with the eventual decision to separate off the military aspect of scientific work from the NRC's normal mode of responsibility. As Mackenzie explained much later, the Council felt strongly that no one body could effectively serve as advisers on both civil and military scientific research policies, 'as the objectives were often diametrically opposed.'[44] Decidedly, Steacie was more than content now to leave Canada's future role in the important task of international control of atomic energy in the capable hands of his former chief General McNaughton, who at that very moment (14 June 1946) was concerning himself with the matter in New York. Steacie may have felt McNaughton's methods not entirely conducive to running a scientific laboratory, but in the political and diplomatic arena into which science was now forced, no better figure existed.[45] Since the country's interests were in good hands, Steacie felt no need to involve himself any further in the issue; besides he was anxious to get on with his real job as director of chemistry in Ottawa. One of the negative aspects of the war had been that, though the application of science played a vital role, science itself, in the form of fundamental understanding of nature and research, was driven into total obscurity. On later reflection, Steacie saw the benefits to science of those years of tribulations in another light: Canada had been forced to enter the world arena and play a role in international co-operation; it had been co-operation with qualification, but through the experience Canada was better prepared for the future. Still, to his mind, something had to be done to restore the balance between the utilitarianism of science and the dearth of fundamental ideas, since it was from those ideas, after all, that the applications of science came.

But if Steacie's thoughts were distinctly directed away from atomic and nuclear issues, the peculiarities of these issues refused to leave him. No sooner had peace been declared when, in September 1945, news of the revelations of the Russian defector, Igor Gouzenko, became known in official circles in Ottawa. One of Gouzenko's disclosures was that the Montreal laboratory and Chalk River security had clearly been breached, from within. Team physicist Alan Nunn May was accused of having given the Russians microscopic amounts of fissionable material, thus perhaps speeding Russian entry into the nuclear business.[46] Steacie's role as senior Canadian administrator in the whole nuclear undertaking ensured that he would have been privy to this confidential information on Gouzenko's revelations. Most likely, C.J. and Steacie locked themselves in conference and retrospective analysis of recent events having bearing on this unexpected development.

While Prime Minister Mackenzie King deliberated on how to handle the Gouzenko affair, May was permitted to take up his readership in physics at King's College, London. It was to be a short posting. An indiscreet newspaper columnist in February the following year ended May's temporary freedom, opening another chapter in the atomic and nuclear story. Canadians now found themselves the centre of world attention. Steacie himself came to entertain grave misgivings about general public intelligence and learnt to be wary of the sensationalists among the press corps, doubts which were to be fortified in later episodes and brushes with both.

But, as much as he may have eschewed the whole business now filling the headlines, it was a situation from which Steacie's name could not easily be extricated. An official inquiry was now in session. When the commissioner's report eventually reached Steacie, a few things were immediately obvious. However the messages had been passed to the Russians, his role at Montreal had been seen chiefly as that of a committee co-ordinator; the information leaked was not at all accurate and the spelling was not proficient either. The telegram to Moscow, dated some time in April 1945, read: 'The Professor advised that the director of the National Chemical Research Committee Stacey told him about the new plant under construction; Pilot Plant at Grandmere, in the Province of Quebec.'[47] The 'Professor' referred to was Raymond Boyer, a native of Montreal who had been appointed to McGill's department of chemistry in 1942 and was undoubtedly well known to Steacie. The garbled message, relayed several times before reaching the Russians, was totally misleading. The plant was not, of course, at Grand' Mère, but Chalk River. Boyer was not involved there, but with the development of the explosive RDX which was entering production at Shawinigan Falls, some miles from Grand' Mère. The remainder of the message, that engineers were being recruited from McGill, that uranium may be used for filling bombs, that the Americans had undertaken wide research and invested $660 million in the business, was of doubtful relevance.[48] But it had all come from a conversation at McGill between Steacie, Boyer, and Carl Winkler, a professor of chemistry.

The subsequent trials were an embarrassment to the administration at McGill, which found itself labelled a hotbed of radicalism and, worse, a centre of 'ideological indoctrination.' When the excitement abated and controversies dispersed, McGill found its reputation of scholastic integrity untarnished. It retained its democratic autonomy, suspending Boyer until after his trial and conviction, and reinstated him to a new post after his sentence had been completed. Throughout all this, the man who suffered most, Boyer aside, was probably Otto Maass. As director of the chemistry department at McGill he felt an overwhelming, though unjustified, sense of responsibility; as an ardent patriot his sensibilities were outraged by Boyer's action. He never did forgive Boyer.[49] As for Steacie, for the time being he kept his own council.

Over the next decade, he was to become further embroiled in the matter of atomic secrets and to find himself labelled an atomic scientist, an occupation to which he certainly never laid claim. His role in the whole affair had been that of pragmatic problem solver, and his activities, though hectic, rarely related to the intricacies of atomic research itself. If his scientific curiosity had been at all aroused, it was only by the faint possibility that all those new radiation studies might have some potential application in chemical kinetics. In 1946, in the midst of these public perturbances, Steacie was more than glad to leave the worries and strains of a problem born out of the war behind. With relief, gleeful at the 'fun' that lay ahead, Steacie returned with considerable relish to some real scientific work, his reaction kinetics.

7

Apprentice in Statesmanship

At the National Research Council the serious business of post-war planning began formally around the fall of 1943. That September there had been the first Review Committee meeting devoted to assessing the future rather than the usual task of reviewing the past. In November C.J. Mackenzie prepared a memo on these developments for C.D. Howe, who wanted to be kept informed.[1] A few months later Howe was named minister of the new Department of Reconstruction and chairman of the Privy Council Committee on Scientific and Industrial Research (PCCSIR), the body to which NRC was responsible. Mackenzie was formally appointed president of NRC. In announcing the new appointments, the prime minister indicated that research was considered to be a vital part of the nation's planning and would be extended and more liberally supported in the post-war period.

Like most people at the Council, Steacie had been absorbed by the needs of war and the increasing demand at the Montreal laboratories. In the fall of 1945, war problems resolved, it was time to consolidate the pressing needs of the future, at the laboratory level. Steacie began his post-war administration by reorganizing the entire divisional operation, including remnants from the war. Some pruning, he decided, was desirable. A few members of staff appeared to have lost their appetite for creative thinking, something Steacie felt was imperative to good work. A lot of communicating behind the scenes, a lot of time, and considerable effort later, Steacie emerged with lucrative offers from industry and other enterprises for these staff. Those who remained did so because they wanted to do science and for no other reason.

Some other aspects of organizing science had never strayed far from his thoughts. As division head, Steacie was compelled to give the whole issue of future direction deep consideration, and most certainly had discussed his increasingly strong views with Mackenzie before pronouncing them to the Review Committee. By this time, the novel decision to move the more industrially oriented section of chemistry to another building was already at the planning stage. During the whole of the past six years some 90 per cent of all endeavours in his division had been devoted to war research. Now Steacie strongly urged that fundamental investigations should be stepped up, and he was already in the process of looking for good men. On this occasion, Steacie expressed an opinion which would become firmly established over the years ahead and leave an imprint which would outlive him. The reputation of the National Research Council, he told the Review Committee, depended largely upon the quality of the fundamental work it produced. The acute shortage of scientists posed a huge problem, though hardly a new one. Lack of qualified personnel had been noted everywhere during the decade following the First World War, but after this most recent conflict, the need for both scientific training and the final products had multiplied manifold and would remain apparently insatiable for a period stretching beyond Steacie's lifetime. As usual, Canada was affected by events south of the border. What well-trained scientists there were at home had become a precious resource for the United States, which, about to launch its own massive expansion in scientific programs, was scouting for talent. Such competition required counter-action, promotion, and better facilities to encourage the scientists to stay. The present situation, Steacie told the committee, was 'very serious and difficult.'[2] Instead of the 10 per cent pure research of the war years, Steacie suggested an increase to 35 per cent of the staff's time. To this end, Steacie formally suggested for the first time the idea of having post-doctoral fellows join the Council's staff on two-year appointments. This innovation, as we shall see, became a major force in the post-war expansion of science in Canada and reached out to encompass much of the world.

The rapid changes envisaged by Steacie for the division were to result in a special subcommittee, culled from the Chemical Institute of Canada and Council members representing universities; Steacie's sweeping reforms were not to go ahead without careful scrutiny by representatives of academia and industry. Both Steacie and Cambron (the designated head of the applied chemistry division now in the process of creation) discussed with this committee the questions which would have deep significance in the life of a national laboratory – 'the balance between long-term and short-term projects on the one hand and between fundamental and applied research on the other.'³ The men were conscientious. The committee of five first perused the mandate of the original Council Act and the recent (1946) amendment before proceeding to pronounce judgment. There were at this first meeting two policy directives of note. The first was that the committee felt strongly that applied chemistry's function was to deal with long-term projects connected with the nation's resources and that service-type research for either industry or individuals be kept at a minimum. Secondly, Steacie himself argued that whilst he concurred with these objectives as major functions, a secondary, but worthwhile one was some short-term studies on behalf of Canadian manufacturers not able to carry out the work themselves. Such functions, he felt, would serve both industry and the NRC by keeping each other well informed.⁴

With the return to peacetime conditions the country as a whole, and science in particular, was poised to take off in a spectacular fashion. It was now the second half of the forties, the war had ended with a colossal boost for science, Canada was looking forward to a bright future. Within the inner circle at NRC, Steacie had proved himself competent, highly so. He had directed with conspicuous success a large variety of wartime chemical activities, played an important role in Canada's nuclear program, and to some degree kept up the impetus of his own scientific work. Conditions within the Council itself were good, prospects bright, research highly exciting, especially since shedding a lot of the adminstrative slog. There was no apparent reason why Steacie should not have felt totally satisfied with his lot.

Yet he was having doubts. Much of his experience of the war years, with all that organization, paper work, and the added tension of secrecy, had not been to his liking. It took him away from the thing he loved best – chemical research. When, early in 1946, a major university offered him a job, he went to Mackenzie and told him he was tempted. C.J. was, above all, a careful assessor of his fellow men. Now that the war was over and so much effort was being put into meeting the future, capable, efficient, and decisive people were just what the NRC needed. Steacie had to stay, though certainly not by coercion; he knew it would not work. He knew perfectly well that Steacie's dislike for paper work was a major factor. 'You have forgotten,' C.J. reminded his younger colleague, 'a lot of things that you didn't have when you were at McGill, and you have forgotten how difficult it was to get a test tube.' Even for a piddling item, he reminded Steacie, it was necessary to get through a requisition; and wasn't that also administration? Departmental meetings? Departmental politics? Steacie had to admit the negative facets of academia. 'Well,' Steacie conceded, 'if this organization will continue to run the way it is running now, I think I would like to stay, but if it is going to turn into [an] administration and [become] impersonal, I don't think I'm interested.'[5] C.J. told Steacie to go away and think about the issues at stake and to tell him the answer in a week. The next day, Steacie was back. He had decided that if C.J. would stay on as president, he would remain.

From that time on, Steacie determined to put an academic career out of his mind and concentrate on affairs at the Council. On 1 October 1946 his term as director of the Division of Chemistry, which expired in July, was extended for another seven years. But within four years, his career would take a different path from that anticipated.

Steacie organized his department's affairs with considerable flair and adopted a uniquely personal approach in solving administrative difficulties. Where others saw an impasse, Steacie merely took a detour. When in 1948 F.R. Terroux at the Radiation Laboratory of McGill wrote to Steacie requesting a quantity of heavy water or, better still, deuterium for their Wilson Cloud Chamber experiments, the latter wrote back cheerfully, eager to help a fellow scientist. 'We have lots of D_2O and can easily let you have 50 cc or so. Also we make our own D_2 [deuterium] and compress it into small cylinders. I'll be glad to let you have the water and the gas if you will let me know the D-concentration necessary.' It was, after all, only a small quantity, but Terroux was worried about the stringency of regulations concerning all fissionable material and heavy water. Should he obtain permission from higher authorities – General Groves, or the Atomic Energy Control Board? Steacie allayed these anxieties with his usual confidence, and a quick solution. 'There is no difficulty about where it came from. Just say you used it, and I'm sure no one will ask where you got it. If they do, tell them I gave it to you.' In case this did not work, 'a simple way to fix that problem would be to write to Keys [Dr David Keys was responsible for the atomic energy project at Chalk River] and ask for 50 cc of water. If you get it you have it as well as what you get from us, and it is all above board.'[6]

By contrast, the physics division appeared, at the beginning of the post-war era, to be lacking in lustre, not to say direction. This was the problem on which C.J. frequently conferred with Steacie. To C.J.'s thinking, 'this division needs a very strong head; it's going nowhere, if you just promote a man who's a senior it'll still be divided; so we've got to go out and get a first-class head, because we want to build up a strong organization.'[7] An opportunity soon offered itself in the shape of

Gerhard Herzberg, and Steacie did not wait for C.J.'s approval, who was out of town anyway, before he jumped.

Herzberg had left Germany in 1935 and found refuge at the University of Saskatchewan. There he had established a reputation for original thinking in the new field of molecular spectroscopy; but, inevitably, expanding research demanded improved apparatus. So in 1945 Herzberg had joined the Yerkes Observatory of the University of Chicago. At his new, though still modest, laboratory, work went well and interest in the improved techniques for studying atomic and molecular spectra accelerated. Still, it was primarily an observatory, and experimental spectroscopy, no matter how original, had to remain a sideline. Besides, at another part of the Chicago campus Mullikan was already operating a large spectroscopic laboratory. The working conditions were pleasant enough, everyone was pleased to share their enthusiasm, and there was a stimulating atmosphere. Privately, however, the Herzberg family did not find the environment of the countryside around Yerkes nearly as congenial as their memories of Canada. More than one Canadian friend knew of their dilemma, and Herzberg had kept in touch with a former colleague, John Spinks. Steacie, on hearing all this in January 1947, characteristically came straight to the point. News was that 'you wished you were back in Canada. Is there any possibility that you would be willing to work here if a suitable offer could be made?' Steacie told Herzberg that a very good group now existed in pure chemistry and 'it is absolutely essential that we do likewise in physics, and it would be of inestimable value to us if we could get you here ... I hope you will excuse my writing to you on the basis of a mere rumour, but I felt that we shouldn't pass up the opportunity of there was the slightest chance of your being interested.'[8]

On his return, Mackenzie took up correspondence where Steacie had left off and wrote to Herzberg, stressing the situation at the NRC.

We are proposing to set up in the Division of Physics a branch which will concern itself almost exclusively with fundamental research and will not have any responsibility in connection with applied work other than the normal informal consultations which would take place between active scientists in the same institution ... you would not be responsible for any administrative work and you would have, within very broad terms, complete freedom of research problems. Dr. Steacie in his Division of Chemistry is operating under a similar scheme ... our philosophy is that in every Division there should be a modest group of scientists who are essentially engaged in fundamental work which is in no way directly focussed on practical problems. We feel that not only will such a group make valuable contributions to science but that in addition the presence of such activity will maintain higher standards in the other sections of a more applied nature and generally stimulate scientific work in the Division.[9]

These were Steacie's views precisely. Herzberg officially joined the NRC in the newly created position of principal research officer that same year. C.J. was well pleased with this success in enticing the gifted scientist to Ottawa: 'Everyone concerned in all ranks is simply delighted,' he told Herzberg. There remained, however, the lack of a divisional director when Boyle retired late in 1948 – resolved in one stroke by elevating the newly appointed research officer to the post. Herzberg's acceptance was not without misgivings, reassured only by the promise that his 'administrative duties [would] be kept to a minimum.'[10] Herzberg's long years at the NRC would bring mutual benefits, culminating in the Nobel Prize in 1971 – an honour which would have pleased Steacie greatly.

These and other developments slowly moulded the stature and scientific reputation of a rapidly growing institution. Mackenzie's faith in Steacie's judgment strengthened with the years; much later he was to recall: 'I have always thought that Dr. Steacie's advice and persuasions were the clinching factors in Herzberg's decision, as he probably became convinced that, if the conditions of work were agreeable to Steacie, he could accept our offer with confidence.'[11]

The problems of running an organization such as the NRC did not, however, hinge solely on acquiring talented divisional directors. Expanding also meant defining goals, accepting new challenges and more urgent planning, increased manpower, much success, and even a few failures. All this took time. Old equipment was replaced by new, inadequate space was countered by new plans; budgets were increased, new divisions added, decentralization begun. When an organization becomes older, so inevitably do its staff. How could its members remain creative and stimulated? In terms of work output, an environment of impermanence, improvisation, and change had long been observed to be necessary, the very anthithesis to a large settled organization within an established hierarchy. C.J. had observed that many large scientific organizations in other countries, particularly when government-run, suffered during periods of crises. In war they were not efficient and could not quickly adjust. What could be done to stop this happening at the NRC, as yet a young institution but subject to the effects of time, like everything else?

Mackenzie laboured over these problems with Steacie. When Steacie had been poised at the beginning of his own scientific career, he had relished the experience of spending time in a fresh environment, under the guidance of a leader in the field. He had travelled to centres of excellence in Germany and England, seen new places, met new people, and benefited from the challenge of learning different techniques and modes of thought. It had been invigorating, a turning-point in his own intellectual development, and he had returned stimulated. Where could this experience best be obtained but with people who already possessed it? And by providing the guidance, facilities, and the means by way of a research fellowship, young post-doctoral workers could be free to develop their own creative thinking and test their abilities. Could this idea be repeated on a larger scale? The training to the doctoral level was a slow process and a good achievement, but the newly acquired knowledge needed time to mature: a fellowship would have the advantage of giving just such an opportunity to the newly qualified scientist, to gain experience; in return more mature members could share their knowledge and the benefit could only be mutual. The principle was certainly sound. Steacie mulled over the idea and discussed it with a few sympathetic listeners, including visiting scientist Frederick Dainton one hot July evening in 1946, softened by the imbibing of whisky. Even in the harsher light of day, the scheme appeared practicable and Steacie resolved to see C.J. as soon as possible.

It proved easy enough to sell the idea of post-doctoral fellowships to Mackenzie. C.J. could see all the advantages of this scheme; for a start it would solve the immediate and pressing problem of staffing an organization like the NRC, now geared for post-war expansion. Initially the idea was not overly ambitious, confined merely 'to encourage the movement of young Canadian scientists eastward and young British scientists westward.'[12] The Review Committee, when tested with this scheme in the summer of 1946, responded warmly, seeing it as training for graduates from Canadian universities, thus reversing the current trend of their going abroad. 'It is understood that such a plan will not make for competition with the universities, but on the contrary, form a pool from which the universities may draw men to the ultimate advantage of all concerned.'[13] In the long run, the plan proved more consequential than even C.J. could have predicted, thrusting the NRC into the rapidly evolving internationalism of science.

Since it was largely Steacie's idea, C.J. told him to handle it. That meant convincing the Treasury Board of the wisdom of the scheme, a task not as straightforward as convincing C.J., even though fellows were to be paid out of the same salary budget as the NRC staff. Steacie succeeded, with a little diplomacy and considerable patience. It was one of the first lessons in government procedure and would stand him in good stead. All the same, the plan was some time in the making.

By 1948 all was approved and formally set to go; there were fourteen arrivals that year, seven from Britain, four Canadians, two from the Netherlands, and one from Denmark.[14] From the beginning, interest was high from all quarters and demand for places even higher: 'the 20 application forms you sent me will be totally inadequate,' NRC liaison officer R.J. Brearley wrote from London in April that year, requesting C.J. to send him at least fifty additional forms by return mail.[15] Fellows were dispersed in different disciplines – to Biology, Physics, and the Prairie Regional Laboratory in Saskatoon – but Chemistry for a long time took the lead. Steacie, not surprisingly, took a personal interest in all these developments.

When the program was expanded to include fellowships to be held at Canadian universities, a much greater expenditure of NRC funds was required. Steacie was again to be negotiator. As remembered by Leo Marion, by then a close colleague, there were hurdles even before approaching Treasury Board. When Steacie consulted C.D. Howe, the latter replied that if he wanted to sell the idea to a higher authority, Prime Minister Louis St Laurent, Howe would make the appointment.[16] Whatever the sequence of events, Steacie extracted a promise of half a million dollars for the scheme and, equally important, official blessing.

For the fellows who applied and were successful, the experience in Ottawa, almost without exception, was to remain a highlight in their memories throughout their lives. As director, Steacie's treatment of fellows attached to his unit was not unlike that which he had experienced under Bonhoeffer in Frankfurt, and Allmand in London. The original chemical problem to be investigated usually originated with Steacie; after that, the worker was left to himself, with just occasional discussions to see how things were going. Steacie knew well the difficulties many new scientists had in putting down their results coherently for publication. Invariably his advice was 'boil it down, keep to the essentials.' If one experiment was not proving successful, Steacie was not averse to change or suggestions; the subject of gas kinetics was proving endlessly fascinating and there was no shortage of problems. Fellows consistently worked long hours for the love of it, revelling in the abundance of laboratory equipment, superb by any standards, an incredible sight for many who came over from war-impoverished countries such as England. The passing of three decades has not dulled the memory of these pioneering fellows; praises still abound for NRC's glass-blower, George Ensell, who could do anything they demanded, thus greatly easing the experimenter's life. Often the lights in the Sussex Drive building blazed well into the night, the enthusiastic workers having returned for another spell after dinner. Many voiced their good fortune in being paid a handsome stipend for doing work they thoroughly enjoyed, in a stimulating atmosphere. The labours of these long hours, for the most part, were greatly rewarded with a flow of publications bearing their names alongside Steacie's, now an internationally established figure in the field of kinetics. It was a terrific start for anyone contemplating a scientific career. Steacie expected only good work, but was not a slave-driver by any means; none of the Herr Direktor stuff, as he had put it a decade earlier. Canadian winters were novel experiences for most visitors and Steacie encouraged them to take time off to taste the pleasures of skiing, or at least falling, in the snow. And after a hard day's sport on the slopes, the men could be found, as likely as not, pursuing more chemical sports in the laboratory.[17]

For the fellows arriving from Europe the salary was princely, around $3,000 to $3,500, a vast improvement to the doctoral stipend of about £500 to £600 awarded by universities such as Cambridge, from which several had just come. An excellent dinner at the Château Laurier then cost about $1.25. Problems of the kind invariably arising from arrival in a new country were softened by the kindness and efficiency of Kay Densmore, the division's administrative officer. Her job combined the practical, such as seeing that new arrivals had Canadian currency (out of a special emergency fund) and finding them accommodation, to the humanitarian, such as bailing out two (later very distinguished) visitors who had been remanded by the police after being caught breaking into a building to retrieve a coat. As for the men (of a total of nine hundred fellows who joined the chemistry division between 1948 and 1975 when the scheme ended, only sixteen were women), arguably the most privileged scientists in modern times, it was a giddy experience. The impact of coming together in a new environment to pursue science, a discipline in high ascendancy, meeting like-minded colleagues from other countries, other cultures, sometimes the other side of the world, was novel for most. In Ottawa, fellows from Europe and North America met fellows from Japan for the first time. It seemed impossible, somehow ludicrous, that their countries could have been at war. They mostly liked each other, shared many problems, and spoke the same scientific language. It was, as someone remembered, like a scientific United Nations. The majority of them returned to their own countries, a few stayed at the NRC, some migrated to the United States, and more (largely British) spread out to man Canadian universities, as the Review Committee had envisaged, though even they were to be surprised. By the end of 1949, the Committee were agreed unanimously that 'Dr Steacie was to be congratulated on organizing a science centre which has an international reputation growing year by year.'[18]

In the annals of Canadian science, the post-doctoral fellowship program must be considered a resounding success. Recognition and an international reputation would be a major factor; for a country growing in global stature, winning a place in the now elevated field of science was of immeasurable importance. It did as much as anything in that era to convince the high and powerful in the old country that the torch of science was burning brightly in the new world. Successful fellows considered an award at NRC a feather in their new scientific cap and, as NRC's fame spread, the opportunity for jobs, after a year or two in Ottawa, was largely for the asking. For a time, almost every new scientific PHD recipient from Britain was gravitating towards the NRC, though only a few achieved it. This was to leave an indelible mark on science in Canada, and on the lives of individuals from five continents and forty-one countries, not including Canada and the United States. For the institution, the scheme achieved its main purpose: the NRC remained active and vigorous, with new arrivals every year and only regret when parting came. There was present in that era another element; the output of work was substantial because, according to the fellows themselves, the congenial atmosphere was conducive to such achievements, both for the individuals and for the institution to which they belonged.

For the time being Steacie's staffing problems had been resolved, at a period when qualified men were still in acute shortage all around the world. This flow of scientific talent from Europe to North America would not be without serious repercussions in the following decade. Nor would the scientific problems to be resolved become any greater, given the resources available. But something prevailed which distinguished this early period from those to come. The overriding difference appears to reside in something infinitely subtle – a feeling of confidence in meeting challenges, in rising to the occasion; a belief that problems, no matter how difficult, were eminently solvable. These men knew their job was to do good research and resolve what they could of the problems facing science. 'Planning science' meant devising new experimental techniques. The rights and wrongs of this approach did not cross their horizon. For that matter, it did not cloud Mackenzie's view from his president's chair, and even less that of Steacie, for whom that year, 1948, was to be a special one. The repercussions of the post-doctoral fellowship scheme would take a while to be felt, but it all added to the general rising spirit which would mark the Steacie era.

Steacie's experiments on heavy-atom photosensitization, begun in the late thirties, had continued through the early forties, in time snatched from the hectic life of the period.[19] With war problems behind him, Steacie returned in earnest to his research, launching into a series of studies on the rates of hydrogen abstraction by methyl radicals. These radicals, important intermediaries in many organic compounds, could be satisfactorily produced in large quantities by the photolysis of acetone. Initial work to determine the necessary kinetic parameters, the activation energy and frequency factor for the H abstraction by methyl radicals from acetone, itself proved a formidable task. Once achieved, Steacie and his team were able to proceed with detailed studies of the abstraction rate, using large numbers of saturated, unsaturated, and aromatic hydrocarbons. The results showed in a striking way that the rate of abstraction of a vinylic hydrogen atom was considerably less than the rate for an allylic hydrogen in the temperature range studied. Relative rate measurements also showed that hydrogen atoms in the beta or gamma position made negligible contribution to the abstraction process. In unsaturated compounds, the variation in reactivity of primary, secondary, and tertiary hydrogen atoms in the alpha position to the double bond was also well marked.

All these results, the analysis of measurements carried out with every practical care and precision, would win increasingly the respect and confidence of the international chemistry community. There could be no doubt that the growing body of data built up by Steacie and his co-workers was contributing hugely to the important field of chemical kinetics. An image of the mechanisms by which chemical reactions occurred was slowly beginning to emerge.

Nineteen forty-eight would prove to be a fine year for Steacie. The post-doctoral fellows, newly arrived in 1948, had quickly settled in and become assets to the laboratory work. Steacie's monograph on free radicals (*Atomic and Free Radical Reactions*, published in 1946) and the shorter *Free Radical Mechanisms* were now cited as standard works. This steady output from his laboratory was rewarded by international recognition. He received notice of election to the Royal Society of London, undoubtedly the loftiest scientific honour in the Commonwealth, a prize for any budding scientist. To the end of his life, Steacie would reserve his highest admiration for the sentiments in this society's motto, 'Nullius in

Verba.' This encouraged men of science from the earliest days to accept the authority of none in seeking after truth and to remain, on the whole, unmoved by the praise or blame which followed. Steacie's own translation was shorter – 'don't put your trust in words' – which to him also described perfectly the much brandished scientific method.[20]

The Royal Society's election process was long and arduous, a formidable exercise which Steacie himself would be tirelessly practising shortly. It was fortunate that war casualties had not included loss of contacts with European scientists; his thirty publications in that period easily saw to that. Those members who actively canvassed for his election – E.J. Bowen, H.S. Taylor, and E.K. Rideal – knew his work well; C.N. Hinshelwood, H.W. Thompson, and E. Guggenheim warmly welcomed him to their midst, all of which showed, as C.J. had noted a year earlier, that Steacie was 'finding relationships in England very congenial.' The trip there, taken in the autumn of 1947, had been largely to spread the news of the post-doctoral fellowship program but had also provided the opportunity to renew acquaintances and visit laboratories. C.J. was more than casually aware of Steacie's experiences. 'He is being accepted as one of the first of the leading world chemists,' the president had noted approvingly, 'his arrangements for exchange of personnel, scholarships, etc., should give very interesting results.'[21]

Prestige apart, being a fellow of the Royal Society had practical advantages. He could now send scientific papers (his own or those of others) directly to the society for publication, which in his earlier days had to go via H.T. Barnes, for example, one of the few Canadians then belonging to the society. His journey to London in the spring of 1949 for initiation into the society would further consolidate his scientific reputation; he lectured at the University of London and up and down the country and generally enjoyed himself. In the years ahead there would follow many such visits, which helped transcend the usual arm's length form of communication. David Martin, the society's executive officer, became a close associate with whom Steacie chose to share problems and confidences of scientific administration. A warm friendship would grow and last until the end of Steacie's life, once Martin (later Sir David) came to understand and accept Steacie's forthright manner and independence of mind.[22] This became only too evident when Steacie began his own canvassing to elect Canadians to the British society, resorting sometimes to wholly uncharacteristic bouts of nationalism. Exasperation at the slow progress led in 1958 to a personal campaign which won chemist H.S. Taylor's unstinted approval. It was quite in order, in Taylor's opinion, 'to remind the Royal Society that there are candidates for fellowship outside the

Oxford-Cambridge-London triangle. It seems that only in wartime is there an acute consciousness of scientists in the Commonwealth and elsewhere.'[23] If Steacie's efforts for recognition on behalf of Canadian scientists appeared somewhat unseemly, he had the grace to be equally direct when admitting, 'I realize that I am guilty of impertinence.'[24]

Steacie maintained his ties not only to Britain, but also to his former associates in Germany. The war was long over before Steacie learnt of the bitter persecution his teacher Karl Bonhoeffer had experienced at the hands of the Nazis. But in July 1947, Steacie wrote to Bonhoeffer at an address in the American zone. Normally the most unsentimental of men on paper, Steacie was evidently moved: 'I was very sorry to hear of the very bad time you have had, and I hope that your health is improving,' he wrote, and attempted to share news which he knew would cheer Bonhoeffer. 'I am getting back to atomic and free radical reactions, now that the war is over. I am sending a copy of my book which I hope will interest you. I will always be greatly indebted to you for first getting me started in this field.' Steacie's regard for his former mentor was persistent. When contacts were finally established a short time later, Steacie reiterated, 'it was due to you that I first became interested in atomic reactions.'[25] On his subsequent visit to Ottawa Bonhoeffer, though now a changed man in appearance, did not talk in detail of his ordeal under the Nazis, either to Steacie or to Gerhard Herzberg whom he had known from their earlier meeting in Frankfurt. Bonhoeffer's pleasure at Steacie's rapid rise in the chemical echelons was countered by concern over the fate of another former protégé, Karl Geib, who had worked in the Frankfurt lab alongside Steacie. After the war, Steacie had been approached by Gieb – who, like Steacie, had accompanied Bonhoeffer to Leipzig and there found himself, like Bonhoeffer, in the Russian zone – asking about the possibility of working at the NRC. Steacie wrote to the president of council on his behalf, adding that Geib had not been a Nazi sympathizer and in his opinion Geib was 'politically weak and a conformer.' Mystery surrounds the subsequent events. In 1947 Steacie already knew that Geib's 'move to Moscow was not entirely of his own doing.' Further efforts by Steacie to trace Geib appeared fruitless, for Mrs Geib wrote to thank Steacie for his help, the Russians having informed her that her husband was dead.

In 1939, when General McNaughton had been called away at short notice, the Council had been deprived of its president after an interval of

Apprentice in Statesmanship 107

only four years. The sticky situation had been resolved by the rapid appointment and smooth transition of power to C.J. Mackenzie, who kept things efficiently ticking over for the next six years by assuming the mantle of acting president. What was more, up to and including the war years, the NRC possessed no designated personnel for the organization or administration aspects of the Council's affairs, apart from the able S.P. Eagleson, an unbelievable situation viewed by today's standards. During C.J.'s first six years or so of tenure, when the staff was expanding dramatically, day-to-day business and decisions fell upon him, sometimes in consultation with his senior directors, and administrative help was limited to a few efficient and faithful members of the secretarial staff. In 1946 all this changed; the NRC established two new positions of vice-president, one each in scientific and administrative affairs. David A. Keys, a physicist, appointed vice-president (scientific) in charge of the Chalk River project, came from the ranks of academia at McGill where he had taught physics since 1929. Finding a man 'with the right attitude' to take care of administration for the whole of NRC was another matter. Russ Birchard, who was appointed vice-president (administration) in 1947, had been an administrator at General Motors. When first recommended, Birchard was unknown to Mackenzie, who was apparently told that 'everywhere he goes, things get quiet.' That had convinced C.J., but nevertheless he cautioned Birchard, 'to succeed you have got to realize that you are working and making the proper environment for scientists to work [in], and you'll succeed when they come to you with their personal problems and you're a friend of theirs.' This particular philosophy on how scientists and administrators should interact would become a major part of NRC policy for the duration of not only Mackenzie's term, but for the remainder of Steacie's life.

There had never been a tradition at the Council of a stand-in or official assistant to the president, although Boyle had been constantly at Tory's right hand. By the end of the forties the time had come, C.J. decided, for something more formal. Perhaps, too, the growing demands of decision-making made him feel more urgently the need of a second-in-command; he was approaching sixty-two. The choice fell upon Steacie, on whom C.J. later maintained, he had had his eye for a long time. Actually, he was not exaggerating. A memo to Mackenzie from Council member A.R. Gordon in 1948 indicates that discussions along these lines had already been aired. Gordon had suggested, 'entirely independently, make Steacie Scientific Vice-President for the whole council.'[26] In 1950, therefore, following the establishment of an additional vice-president (scientific) position, Steacie assumed the role of vice-president at the NRC. Two and a half decades afterwards, Mackenzie related his conversation with C.D. Howe on the matter: 'I think that Dr. Steacie should be the next President of the Research Council but he's got no experience in administration yet at all, excepting interdepartmental, ... wouldn't it be a good idea to appoint him Vice-President (Scientific) without any administrative responsibilities?' To which C.D. replied, 'This is a good idea.' And so it was done. But if there were no difficulties in doing things this way in those days, there would be plenty later on. It appears that Steacie himself had not been consulted beforehand, for C.J. debated with Howe whether Steacie would actually like the job.[27]

At the time of this appointment, it is doubtful that Steacie had as yet developed any special foibles toward the task labelled organization or administration, apart from the odd ruse he thought up from time to time. The particular body set up to oversee quality of work at the National Research Council, the Review Committee, appears to have consistently approved of his particular style of organization in the chemistry division. It was only much later, when administration had become increasingly demanding both in time and in scope, that Steacie began resorting to methods which quickly became a major component of the Steacie legend around the NRC. Answering mail from the public was one such duty. Generally, Steacie was considerate, civil, and if possible, informative. Now and then, a correspondent received a surprise. 'Dear Sir,' a very long letter addressed to Steacie enunciated pompously one day:

Will you be kind enough to let me know what you think is the most pressing need in education in Canada today?

Now I realize this is a question of some magnitude. I only want the idea that is foremost in your mind when the question is asked.

It might have to do with the teaching of infants or children in public school, the method of teaching the syllabus, physical association, the introduction of play or athletics, the discarding or inclusion of the classics, the cost, the timing of vacations, the role of the government, the final product, the reason why many secondary school children do not get to university. Whether university is a place for top students only or for students as well who could take a university degree in an extra year or so ...

I shall treasure your advice ... it is my obligation to address the Trustees and Boards of education for all Northern Ontario from the head of the Lakes far into the hinterland, in September. I purpose to speak on the Pressing Need in Education.

I am enclosing for your convenience a stamped envelope and a sheet of paper.

The sheet was duly returned with Steacie's one-word reply: 'money.'[28]

In 1950 organization of affairs at NRC did not appear altogether daunting – not far removed from the job he was faced with every day, his first love, chemistry. This involved studying the role of atoms and free radicals in reactions. As these particular radicals had a life span of the order of a millionth of a second, the skill required in sorting out the mess was considerable. By comparison, the admittedly diverse problems of administration did not seem all that intractable. To begin with, C.J. set Steacie the task of taking over responsibility for scholarships and grants-in-aid for universities. This was a wise move and one he knew would greatly interest Steacie, increasing his contacts all across Canada. All the same, when asked to prepare a statement for public relations purposes, Steacie's priorities had clearly not wavered. 'In all peer reports etc.,' he wrote, 'I am very anxious to have emphasized the fact that on becoming Vice-President I *do not* give up the directorship of the Division of Chemistry, or the direction of my own research group in chemical kinetics and photochemistry ... the appointment will not in any way affect my present research activities.'

At the age of forty-nine, now officially vice-president and director of the Division of Chemistry, Steacie took stock of his career and achievements to date. They were considerable and he ended on a note of cheerful self-confidence, though with tongue in cheek. 'It is I think fair to say that I am one of the leading physical chemists in the world, and am probably the leading authority in the field of reactions of atoms and radicals.' On the administrative side, in ten years at the NRC, he counted: the building up of the entire chemistry division 'into an internationally recognized laboratory,' the work of making his own pure chemistry group into an outstanding centre of international importance, and the struggle to make the Montreal laboratory function during the war. As president of the Chemical Institute of Canada the previous year, and a past president of the Royal Society of Canada (Section III), his stamp upon Canadian science was already undeniable. He held honorary D SC degrees from both McMaster and New Brunswick and was an honorary member of the Polish Chemical Society. 'I think it is fair to say,' he appended thoughtfully, 'that I have a reputation for calling a spade a spade.'[29]

Even when the job of vice-president had been added to Steacie's daily schedule, work in the laboratory would not pause. The two years of apprenticeship in scientific administration would in fact prove to be among his most productive, with over twenty publications. Fellows arriving to join the kinetics group expected the labour to be hard but rewarding; they were mostly right. Steacie's enthusiasm helped to make work pleasant, and there were moments of unexpected excitement as arguments raged between themselves and other groups of free radical chemists scattered around the United States, Britain, and a few elsewhere. For Steacie, the nearest rival group in these years would be that of his friend W. Albert Noyes Jr, dean of chemistry at Rochester. The race would be close in many aspects. Steacie's citation for Noyes in 1957, when the latter won the Willard Gibbs medal, could so easily have applied to himself: Noyes's contributions had hugely increased the understanding of kinetics and free radical mechanisms. 'It is almost impossible to imagine,' Steacie generously wrote, 'the state of photochemistry if Noyes' work had not been done.' Noyes also recognized the near parallel of their chemistry careers. The two shared more than just an interest in reaction kinetics; committee work and administration were equally persistent. 'I am thoroughly fed up with being a Dean and want more time for chemistry,' Noyes told Steacie. 'Photochemistry may not be very useful but it is still a lot of fun.'[30]

In search of more fun, the small scattered units would organize a discussion group, beginning in 1952, at the University of California, Los Angeles, aptly named the First Informal Photochemistry Conference. Organized by Francis Blacet, the conference was centred 'deliberately ... around Dr. Steacie as the "drawing card."' The honour demanded in return a series of lectures on free radicals to the chemistry department and invitations to similar exertions at the University of Southern California, Caltech, and the relatively new University of California at Riverside. Steacie's reputation as a leader in his field was no longer in question. There was also his gift of clear exposition of a complex subject concisely presented, which endeared him to his audience. This visit took place in the spring and the Steacies enjoyed the sight of golden poppies in the Mojave Desert and other unique delights of southern California.[31]

The informal nature of these discussions greatly appealed to Steacie, who fumed perceptively at the increasing tendency for science to become professionally organized – as indeed chemistry was by the fifties. In June 1958 he gave vent to his feelings, addressing the Canadian Association of Physicists. Where was such an organization going in the coming decade? Steacie's vision of its fate was distinctly apprehensive. 'Your meetings will be so much bigger than they are now,' he told them darkly, 'that they will be hardly worth coming to, except for the purpose of meeting salesmen.' Worse still, professionalism would rise strongly and the society would become more interested in the status of physicists and surveys than in physics. These premonitions, he assured the physicists, were by no means foolish or unrealistic. Look at what was happening in chemistry. The American Chemical Society already had 85,000 members and would grow.[32]

Advances in the chemistry of free radicals since Steacie began his professional interest had been enormous and their mechanisms subjected frequently to scrutiny. But nothing indicates more clearly the degree of complexity of the subject than Steacie's own appraisal of the problem over the years. In 1952 the prestigious Faraday Society of England, encouraged by Steacie and the generosity of various organizations, took the unprecedented step of holding a general discussion abroad, at the University of Toronto. The NRC contributed some funding, as well as the services of Jim Morrison and Kay Densmore from the chemistry division. No opportunities were lost to involve Canadian industry, which responded generously. At the discussion, devoted to the subject of free radicals, Steacie, who would become president of the Faraday Society a few years later, had to report that progress since the last major discussion on the same subject was by no means a matter for self-congratulations. 'From a pessimistic point of view,' Steacie reflected, 'one might perhaps wonder what we have all been doing during the past eighteen years, since many of the questions asked in 1934 have still to be answered in 1952.' The subject had, admittedly, matured along with the worker but slower. In 1934 the emphasis had been on the existence of radicals. Almost any type of mechanism could be and was postulated. By now the chemist could at least predict the types of reaction likely to be of importance. If the maker of mechanisms was compelled to exercise considerably more self-restraint, this had not deterred the growth of literature on the subject. Steacie's own publications were rapidly approaching the two hundred mark, though a few were devoted to work other than free radicals. The meeting was an auspicious occasion for the development of science in Canada, bringing together large numbers of scientists who had never before visited this part of North America.

In 1953 Steacie was honoured with the Baker Visiting Lectureship at Cornell University, which was accompanied by the handsome fee of $2,500. This would occupy him twice a week for five weeks, together with the journey time between Ottawa and Ithaca, New York, a considerable period of absence. The result of his diligence would be impressed shortly upon the chemistry community at large, in the formidable *Atomic and Free Radical Reactions*, an enlarged edition of his 1946 monograph. This tour de force could be guaranteed to awe new arrivals at the laboratories, not accustomed to working with the leader of a large institution who could also cite references accurately in huge numbers – surely a genius possessed of 'a fabulous photographic and encyclopedic memory.' Steacie, when quizzed, was much more modest about his prowess. It had come, he told his young admirer, from too many boring hours of travelling when he had amused himself by planning his book.[33]

These years of achievement were not without some personal sacrifice. Steacie had long since given up golf, a game which he played exceptionally well. Skiing was always possible in winter, but work remained predominant. Most evenings after supper, Steacie would sit in his favourite armchair and pound out his research papers on a small, much-abused typewriter. This was strictly a two-finger effort, but with so much practice the method became highly successful. The family had adapted to this routine which revolved around father's work, and at night the house was a peaceful place to construct thoughts which daytime activities disallowed.

Into this now familiar pattern of continuous work had intruded one new factor which would greatly increase the family's pleasure over the years. In the depth of the war, they had managed to get away from the turmoil of work by spending a few weekends with friends in a cottage at Grand Lake, a beautiful location north of Ottawa. It was a place to fish and swim and to mess around with boats; it was close enough to home and responsibilities at the NRC. One weekend, with war over, the family packed their tents and returned to explore the delights of this area further. They found a spot where a triangular piece of land jutted out into the lake, with an irresistibly enchanting view. Fortunately it was for lease, later for sale, and the Steacies began an adventure into cottage-building. The first structure was a simple affair, four walls and a roof to provide indoor camping. Learning the art of building and plumbing might be fun and Steacie decided to tackle it himself with help from a local builder. The problem of how to start would be resolved by talking to other residents who lived in professionally built cottages around the lake, and learning their wisdom. Adaptation could be a creative process, and Steacie soon developed a flair for carpentry, constructing everything he could. These were sturdy items and very functional, and if not in the first craftsmen class, it was certain that they would hold together long past an average lifetime. Over the years, a kitchen and a bathroom were added and eventually a large wooden deck built out over the rocks to the water. This haven of delight became more and more a retreat from the pressures of life, absorbing all his leisure hours. Friends from abroad were treated to the informal pleasures of Canadian cottage living. With passing years, Steacie began to make time for an activity he had bypassed in earlier days, watching children grow. At the cottage he could spend long hours with grandchildren, exploring the lake, each learning from the other. And when eventually illness struck and Steacie no longer had the strength to make it to the city to greet visiting scientists and dignitaries, they were invited instead to the cottage he had built, a gesture they would long remember.

Steacie's determination not to allow increased administration to deflect his energy from research was no idle ambition. The Review Committee, surveying the scene in 1950, observed that the photochemistry section was as active as ever and had established an enviable international reputation. The division had thirty-six post-doctoral fellows, mostly fresh from university, adding to the academic atmosphere of the laboratory already instilled upon the division's staff by a series of seminars and lectures on different topics of chemistry, organized by Steacie. No one in his division ought to be totally ignorant of other ongoing research, if he could help it. New fellows arriving at NRC brought an alert and fresh approach to all research, and lessened the possible tendency for any group to rest on its laurels.

In the mean time, the building of the new laboratory for applied chemistry on the Montreal Road site was progressing. When the subcommittee of the Review Committee met in December 1951, a few additional points of interest entered the usual, and by now familiar, approval. The committee noted Steacie's opinion that 'from 15% to 20% of the activities of the Council should be devoted to long-term fundamental research' – in which the chemistry division had by now developed a solid reputation. Of course, the members agreed, such research might suddenly develop practical applications. A case in point would be Steacie's own studies in free radicals, as we shall see. The committee nevertheless appended a caution: 'It is, however, a duty of Council, in studying the activities of the laboratory, as a whole, to see that a proper balance is maintained between applied research (which is the prime responsibility of the National Research Council) and the fundamental variety.' If Steacie was surprised by this interjection in 1951, it was not recorded. In any case, the committee departed with its usual encouraging words: Steacie's division was 'well staffed and well organized,' turning out work of the highest quality, and the philosophy underlying its program was sound.[34] This philosophy – the pursuit of excellence in fundamental scientific research – was one major facet of NRC's policy which would invite attention in the years ahead. Not that it was novel in any way. In fact it had been very much a part of Council's post-war plans, and indeed, similar scenes were taking place in research laboratories elsewhere. Steacie merely gave voice to its importance within the national fabric. But his was a successful voice. It would give rise to both discourse and discord which remain unresolved three decades later.

At the end of the surprising affair which had begun with Gouzenko's revelations, Steacie had not had much time to dwell upon the events in which Canada was now irrevocably involved. General McNaughton was deeply engrossed in these issues at the United Nations, so that was that. But in 1950 Steacie found himself once more having to make public statements in connection with the 'secret and spies' interlude of his wartime responsibilities. This had to do with the trials and tribulations of the scientist Leopold Infeld. Infeld, a gifted lecturer at the University of Toronto, was embroiled in a controversy over an impending journey to his native Poland. When asked for a statement on the matter of Infeld's position as a security risk, Steacie told the press that the scientist had had no opportunity to partake of Canadian hospitality in the form of data on atomic research. 'We want to point out categorically that Dr. Infeld never had anything to do with our atomic energy project.' Neither had he been involved in any project classified as secret. Infeld remained bitter over his experiences at the hands of the Canadian press and decades later commented that if Steacie, as the official voice of the NRC, had only spoken a few months earlier, the whole painful episode and fuss 'over my supposedly taking the secret of the atomic bomb would not have arisen.' 'Why,' he asked, 'didn't these two gentlemen [Mackenzie also made a similar statement] speak up earlier? Perhaps they weren't allowed to or didn't dare, or perhaps they wanted me to leave Canada.'[35]

Whatever Infeld's personal suffering during this political episode of the 1950s, Steacie would be unwavering in his genuine distaste for the secretive aspect of scientific research. And although he would not escape entirely the growing intercourse between science and the public, it was all, in his opinion, a gloomy business.

Meanwhile, no matter how Steacie himself viewed the grind of his non-scientific endeavours, in the eyes of the Council at large, and his peers especially, he was emerging as an exceptionally able and gifted administrator. He understood too well the reluctance of scientists who shied away from the time-consuming, largely uncreative task of the administrator. His ultimate bait in persuading Herzberg to come to Ottawa in 1947 had consisted of the promise that 'you would be coming to do fundamental work on any problem you wish, and that you would not be responsible for any routine work whatsoever. The only administrative work would be to do whatever you could to encourage fundamental work in physics here.'[36] It was all a prelude to a new era. This era, in which Steacie found himself leader of the largest scientific organization in Canada, would coincide with a major development: the expansion of science and, concomitantly, the lack of trained personal required for this expansion and the problem of how to amend this in the quickest way possible.

By now it was evident that Mackenzie and Steacie, though vastly different in their own way, were agreed in their ultimate goals of autonomy and good science in Canada, penultimate to which was excellent work at the NRC. They shared a belief in the fundamental importance of scientific research, not just to a scientific institution but to the future of the nation. Such a philosophy, compounded by a largely united work force, would be sufficient to bind any scientific institution together. And events of the next decade proved to be just so. But in the years following, these same beliefs and phisosophies would be called into question in a rapidly changing social climate, would erupt into a world-wide debate which inevitably embraced Canada, and from which there would be no turning back. Herzberg would find himself in the unlikely role of defender of both science and Steacie.

In the late 1940s and opening years of the 1950s, however, even a wise leader like Mackenzie, much less the younger Steacie, could hardly have foreseen these future events. Besides, it would have made no sense at the time. Canada had emerged with confidence from the war in Europe and turned to look to its own future. Of all the consequences of conflicts, none emerged as strongly as the need for science, which had grown in both size and stature, far beyond expectation. Not just in Canada, but everywhere in the industrialized world, the pursuit was for more excellence in science, the hope of the post-war era. Reconstruction hinged on its success. What had begun as a search by a few individuals down through the centuries out of natural curiosity had become every nation's business. Science was in its ascendancy.

8
Leader in an Age of Certainty

Early in February 1952, with Ottawa still in the grip of a freezing winter, Steacie left for Sydney, Australia, to attend the British Commonwealth Scientific Conference. Wednesday, 13 February brought a night cable, paid for by members of the chemistry division clubbing together – NRC could not pay for instructions which were strictly unofficial:

Come back to Canada via Bond Street to buy white tie and tails. You are the President. Congratulations. [Signed] Chemistry Division.

The official version came a short while later:

Your appointment as President of Council announced today. [Signed] Mackenzie.[1]

The news was not unexpected, but Steacie was naturally pleased that it was at last certain. In 1952 the presidency of the NRC was, without question, the highest scientific office in the land. Steacie knew it was an honour; even so, he received the news with some reservations. 'I'm afraid it is going to be a tough job to live up to C.J.'s example,' he told Otto Maass; but, he added with confidence, 'it should be fun.' The decision to accept the presidency had not been all that automatic. Mackenzie, when appointed president of Atomic Energy of Canada Limited (AECL), the crown corporation to be responsible for the whole Chalk river enterprise, was not enthusiastic at the move. He had built up the NRC to where it now stood, he was proud of it, and he wanted to see his plans fulfilled. But C.D. Howe had requested the move to AECL, so there was little choice but to acquiesce. Besides, he was in his sixty-fourth year, and changes were inevitable. There was much comfort in the way he could at least oversee these changes.

By then, C.J. knew his man very well. Before recommending Steacie for the post he had called him in for some heart-to-heart discussions and repeated some of the things they had already spoken of on Steacie's appointment as vice-president. 'You've got to represent Council,' he told Steacie, which would be different from being vice-president and radically different from being division head. Being responsible for NRC called for making speeches, developing social and political contacts, and above all, making decisions. Attending committee meetings and conferences would become a major way of life. Mackenzie knew full well that Steacie held the firm conviction that committees – especially the large variety – were ineffective when it came to achieving results. Even now, writing from Australia to Otto Maass, who had congratulated him on being appointed the new president, Steacie reported glumly, 'I'm afraid that the Conference is just one committee after another – with none of them really doing anything.'[2]

After briefing Steacie on the presidency, C.J. had told him to go home and think the whole thing over and to let him know in the morning. Steacie went away, but C.J. was not particularly surprised to see him back again in two hours. Decisions, to Steacie, were hurdles to be overcome as rapidly as possible. The secret was to stick by them, once decided. This one had required heart-searching and realistic thinking, but he had had more than a small taste of what the job entailed during his dozen or so years at the NRC and more especially in the war years. What real decisions there were to be made had been faced long before now. The two hours were needed really to thrash out the most difficult of Steacie's personal dilemmas. The pull of his scientific work was irresistible. His fascination for the peculiar antics of how minute fragments of matter could determine the outcome of chemical reactions had never ceased. But after two decades in the field, he knew that it was an area of investigation which required time above all else. And time would be in short supply.

Once the decision to accept the post had been made, C.J. ensured that Steacie was in on most of the preliminary discussions. On 7 February C.J. and Steacie had had a conversation with C.D. Howe and Norman Robertson on 'the organization of the new Company and the Orders in Council that are going through.'[3] But when he left Ottawa for Australia, a great deal remained to be done by C.J. before the leadership takeover. There was the task of soothing ruffled feelings over seniority in the ranks. If anyone was to be promoted from within, on the merits of long service to NRC the lot should have fallen to J.H. Parkin, director of mechanical engineering, who had been with the Council since 1929. Throughout the war he had worked hard and well alongside Mackenzie. His personal expectations remain unknown, although later recollections were not without a faint note of rancour.[4] From within the NRC, the general consensus to Steacie's appointment was delight or, as Eggleston put it, 'there was really no competition for the post.'[5]

Years later, Mackenzie recalled that all had gone exactly according to plan: he went to C.D. Howe and told him that Steacie was the best man for the job. Furthermore, the decision had better be made quickly; otherwise, everybody in the country would be wanting the job. C.J. had recommended that Howe put into the same order-in-council his posting as president of AECL and Steacie's appointment as president of NRC. That way there would be 'no problems at all, and the whole thing would be settled.' Howe agreed. The way in which Howe and Mackenzie did business from time to time had evolved from the circumstances during and immediately after the war. The fact that the Honorary Council was not consulted on that occasion was not breaking any rules. The appointment of the NRC president was, and remains, solely the decision of the government. The cabinet of the day depended upon the minister responsible, C.D. Howe, and in matters relating to science it was Mackenzie's duty to advise Howe. On 12 February, immediately after hearing that cabinet had approved both his appointment to AECL and Steacie's to the NRC presidency, C.J. called the directors of the divisions in for lunch to tell them the news. Out of consideration for his seniority, C.J. saw Parkin before the others, as 'I wanted him to know first.' For an old NRC hand like Parkin, it was not an easy time.

Not surprisingly, on taking on the presidency of AECL, Mackenzie did not sever his ties with the NRC. He continued to be a member of Council, and Steacie insisted that C.J. continue to occupy the president's chamber in the Sussex Drive building, which had been his for over a decade. In the mean time, Council proposed to create the position of chairman for C.J., so that he could 'be of assistance to Dr. Steacie' during the transition period and appear in an official capacity for any emergency. After some reflection, C.J. demurred, for fear that it might set a precedent. In any case, the two men had been working together for so long that the change was not nearly as radical as might have been feared. C.J. was able to give all the necessary assistance without resorting to any formal title, simply by having discussions with Steacie, just as they had been doing for the past ten years.

Steacie had returned to Ottawa in March and began his term as president officially on 1 April 1952. Since C.J. would continue to occupy the official president's quarters, Steacie chose a medium-sized room farther down the hall to serve as his own administrative headquarters. It was a laboratory at present, so a certain amount of renovating was necessary. A little antechamber was added to the secretary's office so that visitors could wait in comfort. And in one corner of the office a spiral staircase led directly down to Steacie's laboratory in the basement. This innovation was designed by Steacie himself and was an almost perfect solution. In years to come, when time permitted and in rare moments of leisure from Council affairs, he would retire to the quieter haven of his laboratory and the intellectual stimulation of chemistry. By this literal escape route he managed to keep his hand in among the glassware and the vacuum pumps long after his official title no longer linked him with chemistry. This quick exit, enabling him to return to the life of an ordinary chemist, was by no means wishful thinking on Steacie's part. Over the next nine years, in addition to the two-volume *Atomic and Free Radical Reactions*, he produced some seventy-two scientific papers. These, of course, were not the result of his efforts alone. They encompassed the work and shared labours of colleagues and post-doctoral fellows alike, who remained a part of his life as leader of NRC. Some of the fellows stayed for a short time and then went to further their work in other places and distant countries; a few stayed a lifetime. Undoubtedly Steacie's most constructive days as a scientist were during the thirties and forties. In the fifties his scientific endeavours took on a different emphasis. From being a bench worker, Steacie became the inspiration for and mentor of a younger generation of photochemists.

With Steacie no longer in charge of the chemistry division, some changes had taken place there. Leo Marion was now head of the pure chemistry division while Ira Puddington watched over the applied division. Steacie delighted in the quick switch of roles, from president of the Council to that of the working scientist at the laboratory bench. He ensured, however, that Leo Marion made all the final decisions regarding pure chemistry's administration, especially those in which he himself was involved.

As leader of the NRC, Steacie ranked as deputy minister and was required to acknowledge the necessity of keeping official confidences. As a scientist, however, Steacie was adamant that scientific knowledge had to be open and accessible. But in this new era, more and more scientific work would involve an element of hands-and-eyes-off variety, stemming partly from increased dependence on government contracts. Industries, too, evolved a higher degree of closed-door research. Universities, however, would continue, in theory at least, to uphold the principles of genuine scholarship – truth for truth's sake, knowledge for the benefit of all mankind. All these facets of post-war society would leave their imprint on the fifties; all would have personal impacts on Steacie.

Canada, *Fortune Magazine* proclaimed in 1952, was a happy land. 'It had riches, it had resources, it had a future.'[6] Not until 1957–58 did any cloud emerge on the horizon, and even then the mood of the nation soon rallied around with resilience. It was difficult not to be enthusiastic about the present and optimistic of tomorrow. The governor general, travelling through the country, cast an appreciative eye on this 'wealthy, booming, optimistic, enterprising country, with its growing millions ... vigorous, able, confident that Canada's century, a little late in starting, is making up for lost time.'[7] In the same year a survey of progress at the NRC (admittedly an internal exercise) concurred that the future would be bright.

Two related phenomena dominated the scene in the post-war decade: the nation's expanding economy, and, for the first time, the accompanying national popularity of science. In the latter case it would be a time of unprecedented expansion. Science had helped to win the war and science promised a new and hopeful future. Science was rarely out of the news and Einstein became as familiar as any major political figure. The United States government's allocation of funds to support science and technology had increased steadily after the Second World War to reach a staggering figure in excess of $15 billion a year by the mid-sixties.[8]

The demand for scientists appeared insatiable and the prices some countries were able and willing to pay to get their services ran very high. The rapid expansion of research and development, particularly in the United States, required enormous numbers of qualified personnel. Companies were relentlessly efficient in their methods of recruiting. Teams of men were sent to scout for scientific talent abroad, starting a process which eventually heightened international tension and prompted a debate in the British Parliament. Newly qualified PH Ds were promised ever better working conditions, modern equipment, and rewarding pay. Generous plans were made to fly these young and new scientific elites to destinations across America, expenses fully paid. Their star was in the ascendency, and for many the greatest difficulty after graduation consisted of choosing the most desirable and lucrative scientific establishment – industrial, academic, or government – all of which demanded their talents.

This was the first generation of career scientists trained specifically with a profession in view, expecting and receiving rewards and recognition. It was a decided break from the past. Traditionally, science appealed only to those with a scientific calling; now demand provided a different kind of call. There was a general feeling, which lasted well into the 1960s, that 'every advanced nation faces a shortage of highly qualified scientists for the indefinite future.'[9] To accomplish more science, more qualified people were needed. But the training of scientists was a long and expensive process – a minimum of six years, almost invariably closer to seven or more. As is usual when supply cannot meet demand, a flurry of activity erupted among those with resources to redress the balance. But true to form, the majority of scientists, once settled in a location, bent their heads to the task in hand and rarely looked up. This peculiarly science-oriented characteristic was neither new nor transitory; the ability to concentrate on a problem was a first condition in any scientific endeavour. As a consequence, not many working scientists were willing to or capable of entering the political arena. A small number were forced into the public and political eye; few, if any, came out unscathed. What all of this signified was that the image of science was changing and along with this increased popularity would come added dimensions. Once it was recognized that government and science were now irrevocably bound, a new order of things came into being. There would be many repercussions.

At the National Research Council, Steacie found himself the leader of the nation's largest scientific establishment, whose reputation was a personal legacy from C.J. Mackenzie. Steacie's own method of leadership was to evolve only slowly and from experience, but his hopes and aims, his philosophies and beliefs, changed surprisingly little. In part, this was due to circumstances. The nation was in a certain frame of mind, which accepted his incisive form of decision-making as a part of the country's growing development in science. In the years to come, the relationship between science and politics would not be so harmonious nor so generous.

At the beginning of his decade as leader of NRC, Steacie's general optimism for the future of Canadian science coincided with the enthusiasm felt in the rest of the country. The press, too, was enthusiastic, though there was a hint of reservation. The NRC, one commentator decreed, was a flourishing organization which had made 'an immeasurable contribution to Canada's industrial and scientific maturity.' It was 'efficient, respected, useful.' It had won recognition in the scientific world. More impressive to a journalist writing in 1952, this government institution had kept out of trouble in the political world, and its relations with business and industry and the universities were 'almost miraculously free of trouble.'[10] Of the first of these achievements, Steacie was particularly appreciative: 'We haven't had a nasty word from Parliament in 15 years,' he told *Time* magazine a year later. As to the second comment, Steacie hoped to do better than merely stay away from trouble. His job, he told the *Time* reporter, was in the nature of 'missionary work ... to get Canadian industry interested in doing its own research.'[11]

But Steacie's job would entail much more than maintaining good relationships with industry, as he knew well. As chairman of the Honorary Advisory Council, he would also have the invidious task of dividing the NRC's steadily increasing but nevertheless limited bounty between universities, which were all in need of funds and which all craved special attention. Closer to hand, Steacie had to ensure that the Council laboratories were doing good work and seen to be so occupied. Of course, there was a large dose of public relations flavour about the whole thing. Still, Steacie's guiding hand was appreciated, at least as far as his peers were concerned. 'He's a peach,' they told *Saturday Night* solemnly, 'you can argue with him about anything. He's ready to accept advice. He's steadfast. He couldn't be easier to work with.'[12]

Mackenzie unobtrusively watched his successor with a keen but approving eye. He kept to his word and left Steacie to get on with running the Council's affairs; when Steacie came to his office for discussions, he listened. Only rarely did he step up this latent presence to active participation. From time to time, in the early stages, he continued to point the way to his younger colleague. In the fall of 1952 a series of inexplicable explosions caused serious damage to a grain elevator designed and built by C.D. Howe's firm in Port Arthur, a place so much a part of the minister's early history. Howe now naturally put the problem to his friend C.J. Mackenzie; it was clearly a job for the NRC. Steacie responded immediately. After telephoning the minister, he left that night by plane for Port Arthur with Ira Puddington, leaving two others to follow by train. The arrival was timely. The Saskatchewan Wheat Pool had just got around to asking for someone to be sent – and here they were. C.J. was pleased by this entire episode of first-class statesmanship. 'This made a very fine impression,' he wrote later; 'they are going to set up a working party to do what they can. I am sure Mr. Howe will be pleased; this sort of prompt action makes for standing and prestige.'

Despite C.J.'s warnings on the inevitabilities of public life, the new president initially viewed the time-consuming business of speech-writing and speech-making with dismay. But it came as a pleasant surprise to

Steacie to find that the ordeal could be adapted to his personality much more easily than he had expected. Audiences, he discovered, responded well to him and especially to his clear, resonant voice. There was another reason for enjoying the experience. The art of holding an audience's attention was easy to formulate: confidence in the speaker and certainty of what was being communicated. Steacie had matured with science as a daily companion, had been nurtured into leadership within the NRC, had observed the country's progress as consonant with his own. These would be three elements which he utilized above all in the job ahead. In addition, he quickly found public communication required him to be conversant with another aspect of science and technology few scientists had time to heed – its history. History of science was nothing new to Steacie; he had done well in the subject at McGill. By now, much more scholarship had accumulated, but fortunately Steacie found his digestive and retentive powers to be in excellent condition. He collected what information he could from general reading and began to acquire a personal library on the subject. When consulted by the public, Steacie was eager to share his interest. He recommended H. Pledge's *Science since 1500*, Singer's *Short History of Science*, and Abraham Wolfe's comprehensive volumes on the history of science, technology, and philosophy. The five volumes on *A History of Technology*, assembled by Charles Singer, F.J. Holmyard, and A.R. Hall, appealed especially to Steacie's type of mind, particularly the first three volumes which covered prehistory to about 1700 A.D. Such excursions into the past had a certain fascination and ample rewards, for he saw the tortuous path along which his subject, chemistry, and science in general had come. It only made his resolve as spokesman for science all the stronger.

In all, over the next decade, Steacie was to make over seventy public speeches and addresses. His enjoyment of these occasions grew as confidence increased. He chuckled with the audience over a few favourite stories which he liked to repeat from time to time. There was J.B. Conant's account of early American attitudes toward the relative merits of the inventor and the scientist, or, as the current equivalent had it, 'pure' and 'applied' science.

In World War I, President Wilson appointed a Consulting Board to assist the Navy. Thomas Edison was the chairman; his appointment was widely acclaimed by the press – the best brains would now be available for the application of science to Naval problems. The solitary physicist on the board owed his appointment to the fact that Edison, in choosing his fellow board members, had said to the

President, 'We might have one mathematical fellow in case we have to calculate something.'[13]

There were the more serious occasions when he was asked to speak before a less familiar gathering. Steacie would chew over what to say at home, in the evenings or on weekends. It could be trying for those around him as he puzzled over propriety. Dot soon recognized the pattern. After a period of restless pacing, Ned would sit down, write rapidly for an intense period, his large flowing script filling page after page. Then, just as abruptly, he would stop, put his papers away, and never give it another thought. When the audience was more predictable, Steacie fell into the habit of using prompt cards with the merest outlines. He knew what he wanted to say pretty well by then and he rarely made revisions. Public speaking, far from causing any anxiety, was turning out to be a useful exercise; it required a clear perception as to the past, present, and future of science, in Canada and elsewhere, and the NRC's place in this sequence. All were necessary to plan a workable strategy.

Under Steacie's brand of leadership, the NRC would gradually evolve into an institution of a new era. But there was something undeniably idiosyncratic about the general milieu of this organization. The Honorary Advisory Council over which he presided consisted of members drawn mainly from the universities, with representatives of labour and industry. Although the Council reviewed the work of the laboratories and approved the appointment and promotion of staff, it was mainly concerned with external relations such as research grants and scholarships. Administration of the research divisions was Steacie's domain, and he applied to it his philosophy of choosing able people and delegating to them the authority necessary to carry out this work.

Steacie expected directors of divisions to direct. Council policies were discussed, major undertakings were carefully considered, but innovations and the ultimate responsibility rested with one person, rarely with a committee. When it came to problems, Steacie was never slow to get out of his office and over to where the difficulties were, whether in the same building or out at Montreal Road. All these practices made for a chain of direct communication which would bind together the life of an institution, spreading throughout the echelons. Looking back, the minor things evoked the major. Post-doctoral fellows from abroad, where hierarchy was a part of historical process, observed with awe that the janitors' respect for Steacie was as great as that of the directors. Those who had known him as Ned in the early days were not expected to change now that he had merely moved to the president's office. So it was in this air of informality coupled with responsibility that the laboratories organized themselves. Such a combination worked extremely well, provided the incumbents of the laboratory remained largely harmonious. From time to time, human nature raised its customary capacity for conflict. The solution then called for drastic measures. On the whole, however, the years passed with growing recognition for both workers and the laboratories of the Council.

Much of this recognition would depend upon the judgment, expertise, and good will of committees of one type or another; yet, as we noted earlier, Steacie's confidence in this particular process was never far off the ground. 'I am afraid,' he told W.B. Lewis in 1958, 'that I have a firm conviction that no committee ever accomplishes anything worthwhile under any circumstances whatsoever.'[14] If his beliefs were not justified, there was still time to test them. But in any case, there was little hope of getting away from this aspect of institutional life from now on. He would in time take on the chairmanship of diverse external committees, in addition to chairing meetings associated with the National Research Council, as well as activities at both Ottawa and Carleton universities. His services and skills in these quarters would become increasingly sought after. Administration would move ever closer to domination. In this area, Steacie was greatly aided by his vice-president of administration, E.R. Birchard, who had proved capable beyond C.J.'s hopes and expectations. He was innovative; visitors noted on his wall a picture of a tortoise with the caption, 'Consider the tortoise: he makes progress only when his neck is out.'

But the responsibility of day-to-day Council affairs now rested squarely on Steacie's shoulders. The weight of administration was divided about equally between the public functions and the private: meetings with individuals or groups, replying appropriately to correspondence, and the hard decision-making, where rapid thinking and action were imperative. Steacie was in his element in both. Blessed with an ability to see clearly through tissues of tedium to the heart of the problem, his actions were usually rapid and firm, but fair. Now and then, admittedly, they were merely memorable. There was the time when he asked for a pane of cracked glass to be replaced. Regulations, he was told, required the damage to be a critical size before replacement. Picking up a heavy object from his desk, he soon put paid to the offending glass. 'Now you can change it,' he told the startled worker pleasantly.

His reputation as an irrepressible iconoclast of bureaucratic rules and regulations would soon become legend. This combination of communicative and administrative skills, together with the ability to return to the intense concentration required for scientific work, was a gift possessed by few scientists, even the most controversial and brilliant ones. Of J.B.S. Haldane, it has been observed, he was 'so ignorant of anything to do with administration that he did not even know how to call the authorities' attention to the contempt in which he held them.'[15] Even that most erudite, clever, and successful scientist-philosopher of Steacie's generation, Michael Polanyi, found the organization of a single university department manifestly painful. At Manchester, he preferred to leave the task largely to his colleague in the organic chair, Alexander Todd. This, Lord Todd remembers with amusement, was just as well, as his friend's 'administrative technique usually involved issuing on impulse some sweeping directive ... at lunchtime and then spending most of the afternoon seeking subterfuges which he hoped might ensure that the directive (now regarded by him as bad) would in some way become inoperative without his having directly to countermand it.'[16] No such reputation would ever be associated with Steacie. Faced with a given problem, he was much more likely to seek frontal attack. This natural ability for rapid thought and retort on any occasion would win the admiration and confidence of his peers, and incur the animosity of a few others, although from time to time friends would advise against such indiscretions as casually referring to 'fatuous remarks' made by Canadian MPs, or at least suggest softening them.[17]

Surveying the scene from the speaker's rostrum on the last day of 1952, his first year as Council's leader, Steacie had cause to be optimistic. The trend, he told his audience, was toward Canadian self-sufficiency in research. New industries and the expansion of existing plants were contributing perceptively to the amount of applied research undertaken by Canadian firms. It was all part of the country's growth and contributed to the highly positive economy. It was this recent development – renewed emphasis by industry on building up applied research laboratories – which now enabled the NRC itself to take a new approach, 'to broaden its field of work so as to include fundamental studies, especially those having a bearing on problems related to industrial projects.' But Council's main concerns were on course: from studies of metal corrosion to the improvement of lubricating greases; from the causes of rot and decay in textiles to moulds and bacterial decay; from metal behaviour over wide temperatures to improved fog-horns; from classified defence problems to more efficient radar systems; from the impressive high-speed wind-tunnel, capable of creating conditions in excess of the speed of sound, to investigations on the silting of the Fraser River; from testing of models of freighters to developing a small semi-diesel engine suitable for fishing boats.

Even beyond the laboratories in Ottawa, everything loomed bright. The Prairie Regional Laboratory at Saskatoon sought ways in which science could aid the prairie farmer in finding profitable industrial uses for waste or surplus products. In the east was the new Maritime Regional Laboratory at Halifax, which had come into existence only that year, soon after Steacie had taken on the presidency. The opening, he thought, was an honour justifiably deserved by C.J., who demurred. 'No,' he had replied to Steacie's invitation. 'I won't go anywhere where you are for a year. I won't go on the same platform. I won't be seen [with you] anywhere excepting in private... it is easier for everybody concerned.'[18] As had been the case when McNaughton replaced Tory, a certain amount of transition blues might materialize. Mackenzie had accepted with good grace the honour of staying on in the official president's office, but he told Steacie: 'I will see you anytime you want to, but I would rather not be seen in public with you.' It made good sense. When a particularly onerous decision arose, Steacie made his way to the familiar office. Talking matters over helped Steacie see things clearer. Not that Steacie found himself indecisive about major issues for long, but in the art of diplomacy, he knew C.J. possessed the edge by far. All this was temporary, as C.J. later recalled. 'Of course by the time he was a year [as president], nobody could influence [him], could affect his prestige at all, anyway.'

On the domestic scene, the Steacies made a move from their residence on Ottawa's elegant Island Park Drive to the delightful setting of Rockcliffe Village on the northeast of the city. On a piece of land near the end of a wide, tree-line road, they designed and had built a spacious bungalow. This house would be the setting for many a happy and social occasion such as small private dinners, and the unlikely venue of a large noisy gathering of the usually solemn Faraday Society. But mostly, it provided a peaceful haven where tranquillity prevailed, a place to retire to in the evenings after a hectic day. When work permitted, Steacie endeavoured to be home by six or at least for dinner. Before the children had grown up and left home, these had been unusually lively affairs, with much discussion. To settle a point of dissension or doubt, the children would be encouraged to consult books and encyclopedias right at the table. Steacie's philosophy was that knowledge and meat went well together. After dinner, if his schedule permitted, Steacie worked on his chemistry papers or speeches, ensconced in his favourite armchair with his faithful portable typewriter on a small table in front of him. All this was a well-established routine. But more frequently now, some embassy function or visiting dignitary requested the pleasure of the presence of the president of NRC and Mrs Steacie. On these occasions, if he was the first to be ready, Ned sat reading patiently, waiting for Dot. Doubtless, his mind was not always on the social soirée ahead. 'Oh, are you ready?' he invariably asked absent-mindedly, oblivious to the resplendent appearance of his wife. Such moments require the tolerance of a special person. More attuned to the purely pragmatic ways of the laboratory than the rigours of embassy receptions, scientific demeanours were apt to be, as he once pointed out, 'infuriating to wives and relatives when carried over to general life.'[19]

Steacie took up the presidental reins of the NRC at a time and place which fitted him well; bureaucracy existed certainly, but it was still within the realms of reason. Few, if any, situations arose which could not be resolved, he easily saw to that. He liked his job, his men, and the institution. The first may have been somewhat thrust upon him but the others he would have picked, given the choice. As for Ottawa, it was sufficiently small and flexible to have maintained a highly personal core, which might have been impossible and lacking in most capital cities. Not that the job was without drawbacks. Steacie seriously considered returning to the conducive milieu of academia at least once more during these early years as leader. And on more than one occasion American industry placed generous – exceedingly generous – offers in his path. But these were treated less seriously. 'I have the greatest job in the world,' he would tell Dot. Somewhere along the line Steacie resolved that his fate was tied irrevocably to the NRC and to Ottawa. It was a good decision.

9

The Politics of Science

Something must be said at once. The idea that the entire period over which Steacie presided as head of the National Research Council embraced a halcyon interlude in the history of Canadian science seriously invites qualification. Six years into his term of leadership, Steacie wryly observed his fate thus:

> I occupy a middle of the road position as an engineer converted to pure science, as the head of a government organization barred by the British North America Act from any interest in education, but nevertheless a member of the NCCU [National Conference of Canadian Universities], and finally, as a Canadian, accustomed to taking a mean position between the United States and the rest of the Commonwealth.[1]

We should not be misled, however, by Steacie's tone of resignation. He was not prone to allowing inconveniences such as prescribed positions to dissuade him from his course, when he felt it was justified – especially when these conditions were not of his own making. Nevertheless, Steacie's task was ever to tread the fine line between these invidious demands. The fulcrum of this balance would require some shifting from time to time; the fundamental beliefs would not.

It was a time, Steacie perceived, for action. Far from seeing his era as having reached a scientific zenith in Canadian achievements, Steacie believed that the country's task had barely begun. Nearing the end of his term, he maintained still that 'Canada is, after all, only part of the way along the road from the ox-cart to the bulldozer.'[2] This bumpy progress was, to his mind, hardly surprising considering that traditionally the country had made its mark almost exclusively by producing raw materials.

Whatever the historical recourse, where science was concerned the country was only just emerging from the pioneer stage. This was a state of affairs, however, which he held to be entirely corrigible.

Already in the fifties Canada had to contend with a world in which the roles of both science and scientists were rapidly changing. Many of the problems the country had to confront would be related to the rapid acceleration of the first and a chronic lack of the other. In many ways Steacie was the right man to lead the Council at a time when the country had passed through three decades of anxiety, especially over scientific manpower shortage, with still little light at the end of the tunnel. The solution to the problem lay in building up the foundations on which scientific training depended. Steacie's commitment to the importance of universities in Canada's scientific progress would be unshakeable. He would apply all the powers of persuasion at his disposal, and the effects would be radical and long-term. The decade of Steacie's presidency was marked by certain characteristics which require recognition, for they give credibility to the decisions and events which followed. It is all too easy to forget issues which afterwards seem self-evident. Scientific identity, for example. For Steacie's generation and those before, the road to independence – intellectually, scientifically, industrially, and economically – had not clearly emerged; it was more of a widening track. Because of delayed beginnings, Canadian science was forced to catch up and keep up, both at the same time, as Steacie frequently pointed out. If political autonomy had been achieved, it was not yet the same in the domain of intellectual attainments. Steacie's goal in this direction was unwavering: 'It is certainly not anti-American to feel that Canadian universities should be as good as those in the United States. Nor is it disrespectful to Harvard or Oxford or Cambridge to suggest that McGill and Toronto should be great centres of post graduate work ... it is not anti anything to feel that we should have a healthy Canadian science, Canadian art, Canadian literature.'[3] The policies by which Steacie felt these could be achieved, in the domain of science at least, would be his to formulate in the decade ahead.

In creating government institutions such as the National Physical Laboratory in England, the National Bureau of Standards in Washington, and the National Research Council in Ottawa, the original aims of each organization had been the same – to act as custodian of national standards and to carry out work of a long-term nature in the interest of benefiting industry. Here, however, a few facts would quickly become manifest. On the eve of the First World War, the United States and Canada found themselves confronting quite different shortages. Americans faced with dismay their chronic shortage of materials and metals. In contrast, Canada's natural resources had always been gratifyingly plentiful, but scientifically trained men proved very thin on the ground.

To obtain an adequate picture of the state of industrial research throughout the country was not an easy affair. The original Research Council had attempted this at the very beginning, in 1917. Their findings were not encouraging, except to focus attention on the problem and accelerate government action. At the outbreak of the Second World War, McNaughton had sought to repeat the tedious excercise in order to facilitate mobilization of the country's research resources in the event of war. His survey indicated that some 51 people – 5 physicists, 24 chemists, 14 engineers, and 3 'others' – were engaged in research in Canadian industry up to 1940. A more detailed survey carried out jointly by the Dominion Bureau of Statistics, the NRC, the Department of National Defence, and the Department of Mines and Resources indicated a marginally brighter picture in other areas. Government laboratories employed a total of 249 professional research workers, including 7 women. Universities accounted for another 777, covering all the sciences, including technicians and 136 workers in unspecified areas.[4] The glaring lack of development in industry thus filled a dismal canvas on which was superimposed the accompanying lack of financial support for Canadian universities and lost opportunities for Canadian students. Much of the problem centred on the fact that subsidiary industries in Canada depended excessively upon research which was done in their parent companies and transported to Canada for a price. As a corollary, trained Canadian scientists migrated in the opposite direction. What it all amounted to, when war came, was a shortage of scientists in Canada, an ignorance of research and development on the part of the management of Canadian industries – CIL, Steacie pointed out in 1950, 'was probably the world's largest company which did no research' – and totally inadequate experience in military research.

The shortage of Canadian-trained scientists would persist into the post-war decade, thus throwing a shadow over C.D. Howe's ambitious reconstruction program. In 1947 this projected an outlay of some $40 million, but 'because of the continuing shortage of well-trained technical personnel and certain types of equipment, it is doubtful whether the increase in expenditures forecast for 1947 can be realized.' Contingency plans were therefore laid and 'the departments concerned expect to increase the personnel engaged in scientific work by 25% in the current year.' The outlook, however, was not bright: 'In view of the present-day shortage of trained scientists in most fields, it is doubtful whether this target can be accomplished.'[5] The direction of post-war planning was thus not difficult to ascertain. More significantly, the two critical shortages – industrial research and scientific personnel – were merely aspects of one fundamental problem.

The NRC at its inception assumed the task of building up graduate research in universities by means of scholarships and grants. In 1917 scholarships had been awarded to nine grateful recipients. McNaughton, surveying the same manpower shortage some two decades later, with the prospect of what was ahead, found himself 'entirely in agreement with the gentlemen who had preceded me and the original Advisory Council' who were of the opinion that, whatever else changed, scholarships, bursaries, assisted researches, and grants in aid of research in universities would remain the most important feature of the Council's work. This had to be driven forward with all the vigour that could be mustered. Mackenzie, coming straight from the university environment, was even more familiar with the situation. Grants and scholarships awarded to universities by Council, he reflected, were 'a double barrelled operation. The need was for people ... but we didn't have the universities to do it.'[6] It was a question of building up the universities in order to get these results. It would be a long process. At the end of the Second World War, immediately stringencies permitted, Council proposed a major expansion of its grants program. By 1945–46 expenditure on scholarships and grants amounted to $300,000. Four years later, in 1949–50, the figure had reached $1.5 million.[7]

So it was that, as the nation adjusted to the heady experience of post-war prosperity, universities had set their goals on expansion, anticipating as they did the prospect of increased student enrolment in the immediate future. Demographic studies showed an exponential rise in this demand for admissions. In May 1945 Arnold Heeney, clerk of the Privy Council, had addressed the heads of universities with his concern regarding the 'ever increasing demand by Veterans for the university training provided under the Post-discharge Reestablishment Order.' The University of Toronto responded with a general proclamation, 'Research will take an increasingly prominent place in university post war activities,' and recommended a new Department of Scientific Research at Toronto which would offer 'assistance to industry and government departments in scientific research, particularly in the fields of engineering, physics, chemistry and biology.'[8] This had been implemented at the University of Michigan with resounding success.

Since Canada at this point had not developed its own university research system, it was certainly possible to continue its traditional dependence on other countries for input. The idea was a total anathema to Steacie. Many years later, he would remind an audience of such an alternative. 'We would have been electing for a permanent colonial status, for a permanent low educational level, for a permanent standard of living based on such a status, and for a permanent loss of our brightest people by emigration to more developed countries.'[9] In the increasingly complex world of science and government, Steacie would hold firm to the university as 'the keystone of the whole structure.' Its purpose was to train students, to understand increasingly the works of nature and the seeking of knowledge for its own sake. Above all else, Steacie believed it was essential to maintain this freedom for independent investigation in the face of increasing external pressures. If universities were to amount to anything, they had to lead society and not follow it. The university's duty with regard to the community was 'to give society what it needs rather than what it wants.' The logic of Steacie's reasoning was simple: 'I can think of nothing worse than for the university to follow the direction of public opinion as it drifts aimlessly about under the influence of periodic pressures.'[10]

It should not be forgotten that Steacie's generation had only barely escaped the stigma so long attached to the discipline of science. As a late arrival to the hallowed grounds of high academia, it had long been regarded only as a poor cousin to the humanities. Oxford and Cambridge had followed in this comfortable tradition well into the opening decades of the twentieth century. And although enlightened educators at McGill, such as Tory, strove valiantly to change the status quo, even in a young country it had been a slow process. This is not to ignore the fact that, even then, there existed a small band of conspicuous exceptions – 'a few who tarried briefly in Canada like Rutherford, before moving on to better things,' as Steacie remarked.

There were, of course, a number of other goading factors for increased support of scientific research. The Rockefeller Foundation in the United States had, by 1919, agreed to contribute half a million dollars to universities over a five-year period in the form of research fellowships, persuaded to do so by leading scientists of the day, including Rutherford. The gifts since then had been equally magnificent. In Canada, where there were no such private foundations, the NRC had to do its best. From the pitiful figure in 1917, the number of graduate students registered in the sciences by 1953 had climbed to 1,023. From a starting budget of

$14,000 NRC expenditure on university research would reach $2.2 million in the same interval. This unprecedented growth might warrant, under normal circumstances, a degree of praise and self-satisfaction. Not on this occasion. In 1956 Steacie faced a McGill audience with some grim words. 'Today everything in Canada is booming with the exception of the universities. There is no question that the Canadian universities are losing ground, when they should be moving ahead along with everything else in Canada.' Financial support for university research had increased, certainly, and the number of students with it, but some dangers remained. Canada was still dependent upon its scientifically stronger allies, the United States and Britain.

Steacie's proselytization on behalf of science and universities might mistakenly be taken to be a uniquely Canadian phenomenon. Not so. What occurred in the 1950s took place worldwide. It became as serious in Europe, if not more so, as it was in North America. This shortage of scientists became so acute that, in a sense, those qualified became a precious national commodity – so much so that research institutions indulged in something akin to stealing from one another in order to get the person of their choice. This global demand merely aggravated an already well-defined situation in Canada: an acute shortage of scientists and engineers. The expansion of Canadian industry, Steacie anticipated, would cause an increasing demand for technical personnel and, subsequently, an increase in university enrolment would presumably supply the demand, provided that university staff and facilities could be expanded to a sufficient extent.

In reality, the solutions did not emerge either chronologically or even logically, as Steacie had hoped. Canada's immediate shortage was complicated by the presence of its technologically developed neighbour to the south which served as a magnet for Canada's university graduates. As matters stood, the outlook for universities was not too encouraging. Although government funding was rising, so was the cost of increasingly sophisticated equipment, accompanied by the unprecedented expansion of science. But, as usual, Steacie saw nothing complicated in the problem; it was merely a matter of choice 'to give the universities the support they need or not.' The spate of conferences on the subject of manpower shortage did nothing to help; if anything, it confused the issue. It was irresistible to draw an analogy. The problem, after all, was no different from breeding horses. A shortage of ten million horses, Steacie ventured, could be tackled by calling conferences of elderly horses, having societies pass resolutions, and presenting briefs to royal commissions. This would guarantee the accumulation of much information but no horses. Or, much more sensibly to Steacie's way of thinking, you could get on with the job of raising horses, which took time.[11]

In the more private confines of his office Steacie set to work, translating amusing analogies to words and, he hoped, action. It was a topic on which he would interact with the government frequently throughout the 1950s. For those holding the federal purse-strings, the intricacies of budget-balancing could not be reduced to a tale of horses, or any other tall story. Still, it was an opportune time to press for action. The economy, despite anxieties in some quarters, was uncustomarily healthy; there was the $45.8 million budget surplus to prove it. At the Council, Steacie was having more difficulty in performing the same economic act. Given that universities were in a critical phase of expansion, how could one predict each year a reasonable expenditure to satisfy both the Treasury Board and the ever-increasing numbers of requests for funds from universities? It was an onerous task, and one liable, if not treated with extreme diplomacy, to leave both parties unhappy and NRC's good offices badly tarnished. Steacie tackled it with his highly pragmatic and logical mind. He reasoned that all indications were that industrialization in Canada would continue, with more technological development. And war babies would eventually become university entrants, creating increased enrolment. If insufficiently supported, universities might well find themselves floundering with a burgeoning undergraduate population and a consequent neglect of research. As a result, the number of trained workers might decline just when industry needed them most.

All these thoughts Steacie conveyed to J.J. Deutsch, secretary of the Treasury Board, in the summer of 1955. Deutsch was not ignorant of Steacie's strong feelings on the whole subject. The NRC's support of university research over the years had been 'largely responsible for the whole development of Canadian science over the past 35 years,' Steacie had previously pointed out. Evidence certainly bore this out. Without Council's support throughout those years, by means of postgraduate scholarships and grants for indispensable equipment, Canadian graduate schools in the sciences would have faced stagnation. Steacie therefore suggested that instead of doing annual estimates, a projection of the budget over the next five years would stem a few existing headaches: Treasury Board could plan ahead, universities could be told what would be available, and his task would be less that of a juggler. Steacie stuck out for something over $2.25 million for the current year, rising to double this in the next five years. Those experienced in Treasury Board ways were amazed at Steacie's temerity.

But things were changing. It was not Steacie's custom to raise the topic of other government expenditures. If he had, the figure for defence purposes alone would indicate the breadth of this change in under a decade. From around a projected budget of $250 million in 1945 (an increase of almost four times the pre-war figure), it had galloped to $365 million in 1949–50. By the first year of Steacie's presidency, defence figures had soared to $2,100 million.[12] In comparison, the total NRC expenditure for scholarships and research grants from 1917 to 1954 amounted to under $17.25 million. Steacie told Deutsch firmly, 'considering the very large expansion in Canadian technology and the present and potential expansion in industrial research I do not think that the amounts indicated are in any way excessive.' Canada, after all, had two acts to perform: to catch up with older countries where science was well established, and to match the increased effort apparent everywhere.[13]

This appeal to Treasury Board appears not to have registered any long-term response, for the following year Steacie went on the attack again. In the interim Deutsch had been replaced by D.M. Watters, but Steacie's position had altered on only one detail: he upped the forecast for that fiscal year from $2.9 million to almost $3.8 million. As before, he hastened to explain that NRC support was restricted to graduate training and research. Undergraduate activities were provincial matters and Steacie would never forget it. 'Council's grant program has been remarkably free of criticism,' he told Treasury Board's new secretary. This was due, he thought, to the fact that Council itself made the decisions and it was composed of university scientists, who understood such matters. Nevertheless, the figures showed that for each year, more especially since 1952, expenditures for university support had invariably outstripped estimates provided. The figures now facing Watters projected an increase to $6.4 million by 1960–61. As for the current year, 1956, while the increase was large, 39 per cent, 'it is a small amount relative to the contribution which it can make to the technical manpower problem.'[14]

This submission would receive the attention of a higher authority than before, and be scheduled for cabinet discussion in the autumn. Robert Bryce, clerk of the Privy Council and secretary to the cabinet, expressed a few personal reservations on the matter to Watters. Steacie's proposed program, labelled 'Support of University Research and Post Graduate Training by the NRC,' was, Bryce felt, a major policy question. Bryce, well informed on all matters crossing his desk, observed that Steacie now intended the funds to include training of scientists at the postgraduate level. This, in his point of view, was acceptable to the government. The real problem was that mutterings from quarters concerned with the disparity between the ever-increasing funds for science, and those available in other disciplines, were becoming louder. It was time perhaps for the federal government to take substantial action on the issue of funding university education generally.[15] That same year, a few of these concerns were resolved by the creation of the Canada Council. Its locus would be scholarships and grants for the humanities and social sciences, a counterpart to NRC preoccupation with the natural sciences.

But Steacie's days were not filled merely with the problems of university funding. That other segment of NRC's charge – industrial research – was agitating no less loudly for attention.

The anxieties of the business community on how to increase scientific research in Canada had been brought to Steacie's attention fairly early in his presidency. In 1955 Steacie was guest of honour at the discussion organized by the Canadian Council for Economic Studies on scientific research and Canadian industry. The problem was that research in the physical sciences initiated by industry had lagged far behind research in universities or in public research bodies such as the National Research Council. Forty years ago there had been perhaps only three or four physical scientists of first rank in the whole of Canada. Today there were several hundred, but of these, not more than two or three were to be found in industry. Steacie was asked for his opinion, and contributed a few remedies for getting the best results from an industrial research laboratory:
— only first-rate scientists should be hired
— the laboratory should under no circumstances come under direction of plant management
— routine process control tasks and trouble-shooting should be done elsewhere
— efforts should be concentrated on a few of the most relevant and promising research projects
— management must be prepared to contend with substantial time lapse, of as much as five years, before results could be expected
— industrial scientists should be encouraged to maintain close ties with universities, with the National Research Council, with professional societies, and with laboratories abroad.

The industrialists among the guests, however, were firmly of the opinion that 'there has been a certain amount of ill informed discussion in Parliament on the need for the Canadian Government to spend more than it now does on industrial research.' This was, they maintained, unwarranted, and properly a function of industry itself.

Nevertheless, the gathering was told, industry could benefit from various government agencies. Though operated with particular emphasis on assistance to small industries, in actual practice most of such industries did little to utilize the NRC and other facilities available. The industrialists present also voiced a few complaints – competition for manpower, for example. Outstanding young scientists from Canadian universities either take up post-doctorate fellowships with the NRC and then move on to positions in the United States or they go directly abroad. This had long been an area of contention, but it had been amply demonstrated that most Canadians would prefer to remain in the country if research facilities were comparable with those abroad. This informal meeting ended with a warning, and a note of encouragement. Canadian industry must begin to think in longer time spans when dealing with its basic problems. Top management in Canadian industries could plan to rival other small countries during the next decade 'with far stronger research facilities and manpower than ever before, now available both in universities and in government agencies like the National Research Council.'[16]

Steacie listened carefully and provided his personal observations. He had already received some indication of the business community's present philosophy prior to the meeting. One correspondent, W.W. Goforth, wrote that 'many Canadian corporations like Bell Telephone, Northern Electric ... could profit greatly either by retaining some top Canadian scientist as a part-time consultant to screen and sort out the research information received from their U.S. labs ... or by employing a good scientist (he need not be a Canadian) full-time to do this thoroughly.' The best starting-point, Goforth thought, was not an expensive laboratory or new department of research, but one really good scientist, with a sharp pencil, who knew what he was doing and why.[17]

Then there was the industrialist's point of view, which invariably involved finding the most economical and effective way of getting the greatest research and development value for the lowest possible cost. Such a man had four alternatives: carry out his own research and development; get this done by a research institution or university; contract on a continuing basis by patent rights and licence agreements to have this done for him; or employ some combination of all three. The choice for the manufacturer depended upon achieving the greatest value in this respect, and all in the least time possible. Canadian industry could not afford its own research and development on a wide range of products in the same way as its competitors in the United States, Europe, or the United Kingdom could. Thus, 'the tendency is for Canadian industry ... to enter into long-term licence agreements with leading manufacturers in other countries with the full benefit of research and development.'[18] Steacie was left in little doubt of the industrialist's point of view. His own, however, approached matters from a somewhat different vantage point.

Although the historic causes and processes by which Canada had emerged from colonial status had long been recognized, there were some facets of history harder to adjust to than others. Canadian trade policies, for example, had often been predetermined, set elsewhere, to serve the clearly defined requisites of the British Empire as a supplier of primary resources in return for manufactured products. Such an exchange had worked well in the period up to the Second World War. It was the events since which had provided both the material richness of the present and the uncertainty of the future. It had soon become clear, however, that it was Canada's potential as a supplier of vital resources – nickel, aluminum, iron, and uranium – rather than as a performer, that interested the Americans. Europe's position, too, had shifted irrevocably from the security of its former dominance. By the beginning of the 1960s European powers, after millennia of feuding, were closing ranks, for trade at least, with the idea of a European Common Market. These changes would not go unfelt in both industry and government quarters in North America.

Steacie was thus always at pains to remind industrialists that Canada should be aiming for self-sufficiency. True, 'we owe our degree of industrialization and our standard of living mainly to British and American technical competence and research,' but this traditional dependence of Canadian industries on foreign research had in effect resulted in scientific colonialism. The essential question which Steacie posed with increasing urgency was: 'Can we face a *permanent* situation in which we are largely technically dependent on other countries?' His own stand at least was manifestly certain. From the beginning Steacie was adamant that such a situation was not permanently acceptable.[19] His plans for the advancement of Canadian scientific research would eventually find support from an unexpected source.

When the world awoke on 4 October 1957 to the news that the Soviet Union had successfully launched their space satellite Sputnik I, the impact was significant in more ways than one. It was in the middle of the International Geophysical Year, a co-operative exercise between sixty-six nations to study the planet Earth. Already, in the summer of 1955 the United States, comfortable in its advanced position on all scientific matters, had publicly declared its intention to design and lauch unmanned satellites – a pioneering effort, they thought. Sputnik's appearance was not just a Russian scientific and technical triumph, it had pre-empted Western society's reputation for being inveterate pioneers. Response to this technological feat proved more emotional than analytical. In the United States, the scars would be permanent ones.

Steacie, as usual, exhibited calm, level thinking, and common sense. Why should there be surprise at Russia's achievement? The Russians had expended major efforts in science for the past forty years. What did the West expect? 'This attitude that Russia cannot produce anything original has done the West an enormous amount of harm.' The pendulum's swing in just a few weeks had gone from the supposition of Western scientific superiority to the cry that Western science lagged far behind. This, Steacie averred, was nonsense. But Western science, currently still ahead of that in the Soviet Union, need not stay that way. Russia had demonstrated that science and technical development were not the prerogative of any one nation. 'What we need to do is to consider whether we really are interested in doing everything we can to develop Canadian science ... With reasonable support there is no reason why the future should look dark, but we will have to pay for it.'[20]

Whether the Russian satellites of October and November 1957 or perhaps their first ICBM in August, or even the election of a new government in Ottawa that same summer, had any direct influence on the events which faced Steacie the following year is difficult to ascertain. Certainly, no one connected with the sciences would remain unaffected by these events. It was to be expected that universities would react stronger than most. In January 1958 Steacie received a number of letters from university men, old friends of the Council. Their tone and contents must have surprised Steacie, demanding unequivocal changes in the support of scientific research in the country. They recommended that the PCCSIR appoint a national committee of scientists and engineers to examine national policies for basic research, particularly in the area of more adequate funding, and the government's role in supporting research. More radical still, there were suggestions for the organization of a body in Canada similar to the National Science Foundation in the United States.

A prime mover in this flurry of activities was Gordon Shrum, a member of Council as far back as 1943, serving two terms which had ended only in

1956. It was all the more disconcerting for Steacie to learn that, in Shrum's view, it was wrong in principle for the NRC, 'which requires money for competing services to have charge of funds to be distributed to other organizations for the same type of science.' This view was not new to Steacie, but whether he had ever pointed out that university and NRC's own laboratory support derived from separate budgets, not easily interchangeable, is another matter. 'Dr. Steacie has been very sympathetic,' Shrum wrote, 'but if I were in his position, I would take the same stand, that my own organization had to come first.'[21] This, however, was far from doing Steacie justice. He had been quick to realize the significance of the Russian achievement and its impending impact on Western science. In November 1957, he had already made strong representation to the government for increased support for research in universities. Restrictions on funds for the operation of the NRC's own laboratories, however, would hold.[22]

Steacie replied at unaccustomed length to the complaints on the funding of scientific research from various quarters of the country, giving facts and figures. The previous summer of 1957, the new Diefenbaker administration had arrived with a platform dedicated to achieving stringent economy in all departments of government. This, of course, had an impact on the NRC, which was told no increase would be forthcoming for university support – tantamount to a major cut. Steacie pressed on, put a $1 million increase in the estimates, and 'determined to fight the thing to the last ditch.'[23] He even canvassed the government for new university support programs, in the shape of building costs and establishing scientific institutes, to no immediate avail.

It was at about this point that Sputnik had made an appearance, followed promptly by more action from Steacie, who amended the NRC estimates for university support upward by another $1.5 million. The outcome of all this pressure from Steacie was that Treasury Board approved a rise in university support from $3.5 million in 1957 to just under $6 million for the following year. An increase of 70 per cent in one year for any given program would not happen more than once in any lifetime. In a year purportedly devoted to restraint, it was no mean personal achievement for Steacie on behalf of academic scientific research. Yet now the universities appeared highly disaffected.

Actually, there were good reasons. Steacie may have been surprised, but he understood the genuine plight of the university administrators. A sum of $3.5 million did not go far in equipping laboratories across the country with either staff or apparatus. In the mean time, this new and phenomenal financial increase remained quasi-confidential, sitting in government files with no signs of an official announcement. Steacie had tried pressure, but no budget figures could be brought down while the government remained in a minority position. Furthermore, such exchanges between the president and the government were not necessarily communicated to the university community, which knew only of the previous amount and that it was not enough.

As to the argument for a National Science Foundation (NSF), Steacie did a few quick calculations. For the year 1957–58, the American foundation had a budget of $41 million, compared to the NRC figure of $3.5 – effectively equal when taking into account the population of the two countries. With the new figure of $6 million, Council's university funding would, on this basis, outstrip the NSF. As to the suggestion for holding a national conference, here Steacie explained the need for political expediency.

While the Government is so much concerned with the coming election, any such conference would make a very small effect upon them. The only chance of a real effect would be if the Opposition took it up and used it as a political weapon and this would be most unfortunate ... the worst thing that could happen, as far as the Council's advice being respected, would be for us to launch out at the moment in such a way as to embarrass one political party or another.[24]

From the prairies came suggestions of a royal commission on the same subject of university funding; Steacie expressed dissent, giving the same explanations.[25] From Alberta, in February, came even more surprising news of a recent gathering at the university, which had discussed certain documents. These briefs, Steacie learned, 'were designed as direct approaches to the Prime Minister and Cabinet. There was an implied criticism of the NRC's failure to provide more money for capital grants,' more suggestions of a royal commission and for 'some committee of august scientists to advise the Cabinet.' The universities' urgent and major concerns were mainly two: to increase funds particularly for buildings and equipment; and to see if present government aid (scattered between departments such as Agriculture, National Health and Welfare, the Defence Research Board, with NRC the largest provider) could not be centralized under one body – the NRC, the correspondent R.B. Miller suggested. To Miller, Steacie conveyed his intention of bringing the whole issue before Council in March.[26]

Before the proverbial ink had dried, Steacie had formed definite ideas on how to counteract these clamourings 'by a few individuals for a university voice at court,' as Harry Thode, with whom Steacie opted to share his new ideas, put it. Steacie's first thoughts turned to the formation of a new screening committee for grants-in-aid. Yet no criticism had been levelled at the way NRC apportioned its funds, not even by the strong men on Council such as Gordon Shrum. Then again, the clamourers contended that the universities would get more money for science only by a direct approach. To achieve a separate voice some vague and weird organizations were being suggested.

At a time when Council seemed uncommonly beleaguered, Steacie was doubtless gratified to receive Thode's support. 'The National Research Council is everything they suggest in the way of a satisfactory organization and much more,' Thode wrote. Not that Steacie needed reminding, but Council members, appointed by the Privy Council, were after all responsible for broader aspects of Canadian science, for encouraging and supporting both university and government research. In retrospect, industrial research might have been added to this list. But this would not have altered Thode's opinion that 'the National Research Council has the overall picture, the confidence of the Government, and can urge support for the universities without appearing to be another pressure group asking for a handout.' Would inviting the minister to meet with Council once a year during university business sessions be viable? Steacie responded warmly to this suggestion. 'I am all for it. I always invited C.D. Howe but he never came. I had planned to ask Churchill to the cocktail party or dinner, but he is campaigning.'[27]

Still more news of university gatherings continued to arrive. As with the others, their main objective was to obtain increased federal support for scientific research in universities by some means. Their chief concern was made patently clear to Steacie: 'If the various people working in different fields of the physical sciences were not brought together to act cooperatively, that some group might bring undue pressure to bear on the Canadian Government who from political expediency might adopt a policy which is not in the interest of Canadian scientists as a whole.' All these agitations would appear to point, however obliquely, to discontent among university ranks with the National Research Council and with Steacie himself. Not so, the convener of these particular joint university meetings, Leon Katz, explained. It had been an attempt 'to support rather than to criticise and belittle the excellent work which you and the National Research Council are doing for science in Canada.' The accompanying memo stressed that 'all university conducted research owes much to NRC and particularly Dr. Steacie for his untiring effort on its behalf.'[28] These meetings had ended inconclusively as to future action, apart from a general consensus to put their feelings to Steacie at the earliest possible opportunity. Saskatchewan, Alberta, and British Columbia were not well placed geographically where immediate interaction with Ottawa was concerned.

All these uncustomary disturbances by the keepers of traditional ivory towers gave Steacie much to dwell on but he remained generally optimistic. In the midst of these inquisitions on university funding Steacie told a Manitoba chemist who was hoping to begin work on nuclear magnetic resonance that he hoped the capital grants for equipment would be in good shape for the coming year. He encouraged him to apply for a larger grant, which would allow him to purchase the apparatus rather than attempting to construct it himself.

The situation nevertheless gave little cause for complacency. There appeared, openly, sharply, and perhaps for the first time, a considerable hiatus in understanding between government, the Council, and the scientific/academic community. Certainly universities were facing undeniable difficulties in meeting increased demands; the government was likewise occupied. But if the universities were to remain the corner-stone of the whole enterprise for improving Canadian science, something had to be done. Ostensibly, the issue was that universities generally needed more federal support. But there were, Steacie knew, other factors. Part of the discontent rested squarely on the fact that universities felt strongly that their equipment, staff, and salaries should be comparable to those at the Council's own laboratories. Steacie himself agreed entirely with this; to rectify the situation would remain one of his ultimate aims. This disparity between the university and government laboratory was, however, a fact of life in most countries. Unfair competition for staff, a major bone of contention, was a criticism frequently levelled at the NRC by both industry and universities. Steacie regarded this situation ruefully. The development of NRC had always included that nebulous commodity, a creative, contented environment, in which to do science. Scientists flowed in its direction whenever conditions permitted. But taking the current situation in stock, the pure chemistry division, for a start, was operating solely on post-doctorate fellowships for junior people and had hired only one man in the last seven years. This certainly precluded any charges of competition with universities. True, the applied division had taken on more, but few of these were doing pure science. Steacie indulged in a rare bout of introspection. 'It is always easier for a third rate man to get a large salary for extracting manure than it is for a first rate man doing real scientific work. I will admit that we have a few people like Herzberg and myself who are able to get the best of both worlds, although in the last six months I think I have had the worst of both!'

Much more ominous from Steacie's vantage point was the insidious suggestion creeping more frequently into correspondence – the idea of a central department of science. Nothing would destroy the quintessence of his philosophy faster than establishing such a central organism. From this would grow the strangling tentacles of bureaucracy and the death of creativity. This, he felt, would wreck the whole of scientific research in Canada, and NRC's role along with it. Steacie explained his point of view: 'The most vital thing in our whole operation is that we are not a Government department and we have been engaged in a steady war for forty years to avoid this. If a Department of Science was set up with a Minister and a Deputy Minister below him we would then lose all freedom of action [in administering the university program], commence to get political interference with grants and scholarships, [and] bring the Civil Service Commission into our operations.'[29]

There was one obvious solution to the clamouring of the universities for a stronger voice at the court of financial appeals: they should have it. By the end of February 1958 Steacie had already resolved to try to revive the Privy Council Committee on Scientific and Industrial Research. It would be summer, however, before matters could be set in motion. The new minister of trade and commerce, Gordon Churchill, had first to be initiated into the intricacies, and Steacie also took pains to consult Robert Bryce, secretary to the cabinet. All concurred with the idea. It was time for the committee, which had not convened since 1950, to commence business again. 'In the first place,' Steacie wrote J.D. Babbitt, 'it would get away from criticism that there is no coordination at the Cabinet level.' Secondly, the aggrieved academics, and societies such as the Canadian Association of Physicists which had added their voices lately, could be expected to feel much better if they thought their suggestions were being seriously considered by a cabinet committee. The essence of this exercise, Steacie stressed, would be to show 'that general scientific policy and needs could be brought to the attention of a group of Ministers.' It would also be good for the Council. Steacie asked Babbitt, the secretary of the Advisory Panel on Scientific Policy (APSP), to arrange a panel meeting for September to prepare an agenda for the cabinet committee.[30]

But calling a meeting of eight ministers and the secretary of state for external affairs proved a longer exercise than anticipated and was delayed until late in December. Not only was there to be a general review of policy on research and development throughout Canada, but the problems to be debated would also include federal support of university scientific research, and space research.[31] Those aspects of scientific research on which Steacie felt exceptional attention should be focused – the weakness of Canadian industrial research, and support of scientific research in universities – were summarized in a lengthy survey distributed to ministers by the Privy Council Office in preparation for the meeting.

Although the December meeting itself produced no decisions directly related to the universities, their needs had been made part of the renewed debate on the support of research by the federal government.[32]

10

Science and Government: The Heeney Report

To understand the special relationship between the government and science as it stood in Steacie's day, it is necessary to take into account earlier episodes of Canadian history. When it came to doing urgent business, the war years and the decade or so after were special years. In Ottawa, the unofficial way C.J. Mackenzie and C.D. Howe went about their decision-making has been frequently described. It was not that different in the machinery of government on Parliament Hill. Even an appointment as important as Canada's representative to the Soviet Union was, apparently, a casual affair, withal that the man happened to be the best qualified candidate. The post was arranged 'in the casual manner that was common in those days.' On the way to lunch in the Château Laurier cafeteria, Norman Robertson asked Dana Wilgress if he would like to have the job. 'Wilgress said he would and the matter was settled.' In retrospect at least, the period was marked by harmony, 'when influential public servants ... were active promoters of new ideas and approaches that they persuaded their Ministers and the Cabinet to adopt.' The reason, if one is needed, was really quite simple: 'the problems they faced, though no less awesome than those of today, appeared soluble and their confidence in dealing with them was enormous.'[1]

Even more unique to that era, perhaps, was the way scientists, with the fate of nations in their hands, did business with each other. The story of the eminent chemist Hugh Stott Taylor, chairman of the chemistry department at Princeton on the eve of war, was a case in point. Being British and an alien, regulations forbade Taylor being privy to any secrets, into which category war work squarely fell. This meant that Taylor could not enter his own laboratory, a great inconvenience, since Princeton's faculty contained a not insignificant number of scientific luminaries, including Henry Eyring and Eugene Wigner. This was sticky for the Americans, as Taylor also happened to be the best consultant on heavy water that they had. Such potential problems could be resolved gratifyingly quickly when bureaucracy was summarily ignored, as Mackenzie recalled with relish, several decades later. 'Vannevar Bush phoned me, and said, "Can you put Taylor on the Research Council staff at a dollar a year?"' Of course, this is what Mackenzie did. Taylor, now officially working for the NRC, was seconded to his own laboratory at Princeton.[2]

Mackenzie would long be in the process of 'educating' the government to his way of thinking, by a process of reasoning, holding up a succession of NRC successes, and by a timely display of talent. He had felt confident in the government's continuing empathy, all other unforeseen events, such as elections, notwithstanding. He once boasted to Henry Tizard: 'I used to tell Bush and Conant ... that whatever success we were having in Canada was due to the fact that we had a government organization which, while not bureaucratic or dictatorial, was capable of exercising some general oversight, and that we had no doubt that after the war this philosophy would not only persist, but would bring about a strengthening of our system.'[3]

But there had also been some sombre realizations. Few could doubt that the Second World War had made the role of government in science and research a major, permanent – and public – activity. Mackenzie was convinced that, whatever the situation in government support for science before the war, afterwards it would be different. In the United States, before the conflicts began, private industry was spending four dollars on scientific research for every one dollar spent by the government; in the post-war era, C.J. considered that the ratio of expenditures would more likely be 60/40 between the government and private sector. To his way of thinking, under no circumstances could the government be left out of the planning picture and he had already expressed some thoughts along these lines to C.D. Howe. The fragile ties between the organization of science, the NRC, and the government could hardly fail to impress themselves on a man as sensible as Mackenzie. Besides, on formally appointing Mackenzie as president of the Council Prime Minister King himself had been moved to comment on this very matter. From the government's standpoint – the excellent work of the NRC notwithstanding – the prime minister thought the organization of the Council rather loose and badly defined and wondered who had instigated the original arrangement. 'I suggested that it was probably done in a great hurry in the last war and that no one had really put his mind to it,' C.J. had answered diplomatically.[4] In the next few years, while Council had launched energetically into its post-war plans, a figure close to the prime minister was about to put his mind to the matter. As clerk of the Privy Council, A.D.P. Heeney would play an intimate and instrumental role in the organization of government affairs.

Throughout the late 1940s and into the 1950s, the attempt to find a workable mechanism by which science, with all its ramifications, could be co-ordinated would be made by government and scientists alike. The Advisory Panel on Scientific Policy (APSP) was established in June 1949 under C.D. Howe to assist and advise the PCCSIR 'in the formulation and conduct of government scientific policies.'[5] While the cabinet committee met infrequently during the 1950s, the advisory panel met fairly regularly. It had been one of Steacie's first official tasks as president to chair the meeting of this panel, in May 1952. The Massey Commission report, published less than two years earlier, and largely concerned with the arts and letters, had recommended that a survey be carried out of all federal government scientific research activities, an idea which had not met with any great support. Steacie felt strongly at the time that since the commission agreed scientific activity under the federal government 'was reasonably well in hand,' there seemed little advantage and much waste of resources in such an exercise.

This comfortable position would change perceptively over the coming decade. Already by the end of 1952, Canada's increasing status in international affairs at the political level had precipitated a corresponding position in science. Steacie informed the secretary of the APSP in December that year that External Affairs and the NRC had decided to press ahead with a subcommittee to deal with questions concerning international scientific organizations, a rapidly developing presence. But a large number of subsequent panel meetings had laboured exhaustively under a plethora of requests for funds to attend overseas scientific meetings or approval to hold international meetings within Canada. This was all very well and necessary, but it took time from overworked committee men such as Bryce. Could Steacie provide some views or principles that might govern the general approval of such grants, Bryce had asked. The panel, after all, 'cannot be expected to deal with all international bodies which may warrant some assistance from the government in the holding of conferences in Canada.' Steacie agreed wholeheartedly. Surely such matters could well be dealt with simply by correspondence with the secretary of the panel.[6]

Bryce had been appointed to the APSP in 1949 as a representative of the

Treasury Board and was no stranger to the intricacies of NRC's workings. Less than a year before, the tedium of looking into payment for work performed by one federal department or agency at the request of another had landed squarely in his lap. This was a long-standing source of contention. What the advisory panel hoped for was simply a consistent policy. Nothing could have proved less feasible. What Bryce came up with was four pages of complicated reasoning, which he addressed in the first instance to Steacie. As he saw it then, 'research work properly so called, however inspired or originated, should be paid for by the department or agency carrying it out, but routine testing or analytical work, and possibly routine engineering services, should be paid for by the department or agency for which the work is being done.' Of course there would always be exceptions. The complications became increasingly apparent as the reasoning proceeded. It was important to choose a plan which did not entail 'rather arbitrary decisions of judgement,' or, perhaps worse, which 'would be hard to explain as constituting a logical and consistent pattern, particularly over a period of years.'[7]

For his part, Steacie constructed a masterly reply to what was not, he knew, an easy matter for a government finance official to deal with. The trouble was that the real issues were even more complicated than Bryce could have imagined; payment was relatively minor compared to these larger stakes. Harmonious interdepartmental relations were vital but growing ever more tenuous. For a start, an organization like the NRC had to ensure that its funding did not depend too heavily upon the whims of another department or agency. The dangers of this were clearly demonstrated by recent events in the United States which Steacie was closely monitoring. There, the National Bureau of Standards had found itself on a budget 85 per cent of which was provided by the Defense Department, and as a result was having to hold its breath while the secretary of defense debated future policies of its own department. If radically altered, it could, Steacie had pointed out to Bryce, 'destroy the National Bureau of Standards in one motion.' On these grounds alone, Steacie would be ever vigilant in ensuring that no similar threat ever cast its shadow over the Council.

Then there were other disconcerting facets of inter-agency work which had yet to be cleared up. The NRC was, in 1953, carrying the load of a communications branch, a situation which seemed particularly unreasonable, 'for us to pay a half million dollars a year for cipher books for the use of other departments.' It was, moreover, an activity for which 'we have neither interest nor particular sympathy.' And there were several other problems besides.⁸ In fact, the issues raised by this question of interdepartmental agency payment would remain delicate and difficult. Bryce, by the time he finally came to drafting the memorandum for the advisory panel that winter, had had a good taste of what the scientific side of government work involved. And when it came to the whole area of development work, the multitudinous problems raised appeared insurmountable. Bryce had to admit, 'the Department of Finance feels less confident in making a definite proposal.'⁹ It seemed there was no clear-cut solution to this science business to be had anywhere.

In January 1954 Bryce and Steacie met to discuss advisory panel problems over lunch at the Rideau Club. The major discussion, however, had not been about the advisory panel but about its mainstay, the PCCSIR. They had agreed that the amount of business did not seem to warrant meetings of this committee. But for Bryce the matter of the PCCSIR would not necessarily end with lunch. Perhaps the committee itself need not be troubled unnecessarily, but what about its responsibilities under the Research Council Act? 'It seems to me undesirable,' Bryce wrote, 'to have lists of statutory responsibilities that, in fact, are not being discharged at all and which, as I gather you feel and I feel, are impractical.'¹⁰

As matters stood, then, in 1958, the nation's scientific affairs were in the hands of the Privy Council Committee. Advice to this committee could come from either the NRC (that is the Honorary Advisory Council for Scientific and Industrial Research) or the advisory panel. Steacie reasoned thus: while the Council was composed of individual scientists, mostly from universities, and thus capable of keeping the committee informed on broad scientific activities across the country, the panel consisted of senior government officials concerned with departmental activities. By all accounts, the two existing advisory bodies were admirably complementary. The whole state of affairs, it would appear, was highly satisfactory. That it would continue to be so, Steacie felt no real reason to doubt.¹¹ And yet, in many ways, that year would mark the beginning of the end of a special era.

At first sight, the report of the Civil Service Commission, widely known as the Heeney Report, appears as a minor episode in the history of the NRC. Council was far from being the object of this lengthy review of the public service, carried out under the chairmanship of A.D.P. Heeney. In fact, the Council was relegated to an appendix entitled 'Outside Agencies.' However, the report raised a large number of points which later, long after Steacie's death, were to become central issues of controversy. And it is by far the single most carefully documented chapter in Steacie's decade of leadership. No other incident raised so many issues about which Steacie felt so passionate; at no other time was he more adamant or persuasive than in his handling of this event. Those who observed him in action have commented since that if the threats, as Steacie saw them, had been personal, rather than aimed at the NRC, the response would have been negligible. In the mean time, Steacie and Heeney became locked in a battle which had surprisingly bitter overtones; even the press would become involved in a highly unusual way. Most important, the episode illustrates in a very real manner the nebulous and tenuous ties between government and science. It showed clearly the growing separation between science and the government of the day, and also over the mechanisms by which science was accepted or rejected.

Arnold Heeney was no stranger to the activities of the Council, even in C.J. Mackenzie's days. As clerk of the Privy Council, he had been responsibile for drawing up documents between NRC and the AECB for construction of the Chalk River project in 1947. Earlier still, Heeney and C.J. Mackenzie had both accompanied the prime minister to the Quebec conference. A staunch member of the civil service in the Whitehall tradition, Heeney was one of a small number of able men who were gaining considerable power in Ottawa. Like Steacie, Heeney was of Irish descent, born in Montreal. In fact, their early lives held many things in common: each attended and taught at McGill, they were close in age, and they had arrived in Ottawa only a year apart to take up positions which would eventually lead to leadership roles. The differences, however, proved considerable. For Heeney, the deep influence of a religious family background moulded a character of great conscientiousness in the line of duty as he saw it, of rectitude on procedures, and, with maturity, increasingly aware of the hand of God in his affairs. Despite this, Heeney frequently suffered bouts of grave uncertainty.[12] Steacie, in contrast, without guidance from a father and clearly ignoring the religious bent of his mother, had set out on his own intellectual development with no qualms; hesitation over decisions of any nature were practically unknown.

By the mid 1950s, Heeney's and Steacie's paths had often crossed, both professionally and socially. On an earlier official visit to Washington, Steacie had been hosted by Heeney, then ambassador to the United States. At home they were near neighbours in the close Rockcliffe community. It is doubtful, however, if they ever discussed personal philosophies. Certainly their views of contemporary history differed widely. In 1955, at about the same time as Steacie was expressing strong views on the atomic age (nuclear power, he maintained, was merely a natural development in science and civilization), Heeney was struggling with a different set of beliefs. 'I know full well,' Heeney told a group of graduating students in that year, 'that in a sense – a very real sense – the physical existence of the human race is threatened with extinction by the terrible engines of man's own devising ... It would be foolish to underestimate the solid ground for grave anxiety to which we have been brought ... by the prodigies of physical science and by our failure to achieve comparable progress in politics and morals.'[13] Heeney was worried that, having emerged in triumph from the threat of totalitarian regimes, society was now veering to the opposite extreme – a strong desire 'to criticize the inefficiencies of bureaucracy.' Clearly, Heeney cautioned, the dream of political theorists for the day when the organized state would no longer be needed was not yet ready to be realized.

The establishment of an organized state was Heeney's persistent preoccupation. His training had been composed of such matters; his entry to power beginning as principal secretary to the prime minister in 1938 encouraged and nourished it; his recent post as ambassador depended upon it. 'The difficulty,' his mused, 'is to determine at any stage in the development of a free society, the extent and nature of the functions which, in the common interest, it is desirable that a state should exercise.'[14] This, he thought, called for a continuous process of scrutiny.

In November 1957 Heeney, not long returned to Ottawa from Washington to take up the post of chairman of the Civil Service Commission, wrote to Steacie concerning the review of the civil service, already under way: 'In this large undertaking ... the Commission can profit greatly from the views and experience of departments. For this reason I am writing you now to invite your considered views on personnel management in the Service.' The words of Heeney's communiqué which struck home were: 'We wish particularly to have you comment on ... the division of authority as between departments and the control agency (the Civil Service Commission).'[15] Steacie's reaction was to be expected. Since the NRC was not a government department and did not come under the authority of the Civil Service Commission, the questions posed did not arise. There, it appears, the matter stood until December of the following year when the report, *Personnel Administration in the Public Service*, was completed. Controversy erupted and would continue for almost a year. When it finally died down, there remained a breach between science and the government which had not existed previously.

Late in 1958 Steacie departed for an official visit to India, comfortable in the knowledge that the NRC was in no imminent danger from the reforms promised by the Heeney Report. The president had taken the precaution of consulting the chairman of the Civil Service Commission personally regarding the relationship of the report to the Council and, as the vice-president of administration, F.T. Rosser, later recalled, had been assured that there was nothing in the report which would suggest in any way that the NRC should be brought under the control of the Civil Service Commission.

The reform of the civil service was a long-desired project for Heeney and he had proposed a royal commission for this purpose as early as 1945. The process would not be a simple one, and doubtless when he had given Steacie his personal assurance over the non-involvement of the NRC, his mind was preoccupied with much wider issues. This appears surprising, in view of his predilection for correctness of procedure. What was more, it constituted a second error; he had already neglected to distinguish the NRC from a government department. Both mistakes, though in themselves minor, were to prove unfortunate, for Steacie would interpret Heeney's words literally on each occasion.

When Steacie returned to Ottawa, it was to find that all was not well. The recommendations in the report appeared to bode ill for the future of the Council. So, together with Rosser, he issued a lengthy statement on the NRC's position. In particular, they took exception to the opinion expressed in the report that the exemption of the NRC from the Civil Service Act had 'adversely affected the efficiency of the Public Service.' On the contrary, 'the efficiency of the NRC has benefited greatly by being exempted from the provisions of the Act.' When the Council's laboratories were planned in 1924, Parliament had specifically exempted the Council from the Civil Service Act. It was firmly believed even then that if the laboratories were to be successful, staff appointments and promotions should be under the control of the Council itself. This belief, Steacie and Rosser felt, had been thoroughly vindicated during the past thirty-two years of the laboratories' operation.[16]

Because the points raised by Steacie and Rosser embodied the very philosophy by which the NRC had operated since its inception, they were of greater significance than might at first appear. Throughout his term as president, and for some years before, Steacie held a number of criteria as imperative to the successful operation of the NRC, chief among them that the Council must be independent to select, promote, and generally handle its own staff. This was part and parcel of his uncompromising stand on the principle of selecting only outstanding or excellent people for the laboratory. When Council was originally set up in 1917, the staff had been appointed by the Civil Service Commission. This had been revoked by a new act seven years later, which approved the operation of laboratories in addition to the Council's previous activities. 'It was recognized clearly by all concerned,' Steacie and Rosser wrote, 'that the Civil Service Commission was not fitted to meet the problems of staffing with creative scientists a national scientific laboratory.' It was in the public interest to make the National Research Council responsible for the administration of such a highly specialized group. The NRC, Steacie added, was recognized in Canada and internationally as providing an excellent example of what could be achieved in the field of personnel policy and practices by an independent scientific organization managed by scientists. (Actually, this piece of apparent immodesty was soundly based. Under Steacie's term of office, NRC's reputation had reached heights hitherto unknown, both at home and abroad, as his correspondence would increasingly indicate.) Under these circumstances, in his opinion, there appeared no reason why NRC or any other well-run organization should be put under the commission's jurisdiction for the mere sake of uniformity. In scientific research there could be no such thing as job specification. Under the Civil Service Commission, the search was invariably for qualified people to carry out a specific task satisfactorily: in short, the job comes first, and the person best qualified to fit it is selected afterward. By contrast, in scientific research, 'the scientist makes the job,' and the necessary criterion for creative science was that the job had to fit the scientist.

Steacie also pointed out that the Council's policy of appointing scientists for renewable terms of one, two, or three years differed from that of the civil service. It was a well-known fact that creative scientific work was usually done by people in the younger age groups, and it was for this reason that Steacie had instigated the post-doctoral fellowship program. This scheme had encouraged the turnover of junior and intermediate scientists, in the interest of providing trained research workers for Canadian universities and industry, and brought a fresh supply of ideas into the Council organization. Steacie contrasted this with an administration where the experience gained with increasing age was quite desirable.

In point of fact, one of Heeney's objectives in overhauling the Civil Service Commission was to redress the lack of co-ordination and uniformity between the practices of the civil service and the agencies. As Steacie saw it, the staffing problem of an organization like NRC could not fit into the mould of the established civil service. 'The NRC's primary function,' Steacie and Rosser ended,

is to promote scientific and industrial research in Canada. Science is international and to achieve recognition demands a high level of excellence. Since the quality of research depends upon the training and abilities of the people doing it, the personnel policies of a scientific organization are basic to its success or failure.[17]

This ideology would surface as a focal point for later criticism, which saw the NRC under Steacie's leadership as an institution 'built by scientists for scientists,' the very embodiment of élitism.

These sentiments, however, were by no means unique to Steacie. As we saw earlier, A.B. Macallum and H.M. Tory, Steacie's predecessors at the Council, held similar views during their day. But it fell to Steacie to express them with a vehemence not previously known, and to bring the issue into public prominence. It was a sign of changing times that this controversy between the president of NRC and a member of the civil service should make the news columns from Ottawa to Montreal, from Toronto to Edmonton, and from Guelph to Calgary. Few could have guessed, even then, that the controversy would rage on, gathering momentum in the months ahead, fuelled by wide and outspoken support from the press, which regarded the NRC with national pride. The editorial of the *Ottawa Journal* for 7 February 1959 had already expressed the opinion that the recommendations regarding the NRC in the Heeney Report should be thoroughly examined before being approved. It took the view that the Council's rapid and impressive attainment of high international repute rested on its ability to lure to the Council the brightest and most adventurous minds, Canadian and foreign – types whose concern was not so much with security of job as the 'opportunity to think and probe and experiment.' While commending the bulk of the commission's recommendations, the *Journal* warned, 'We must be wary of the tidy mind. It sounds fine to assert that almost everything in public service should be brought under one surveillance, it has a ring of efficiency to it. But it is a generalization to be examined with care and even suspicion.' Steacie's own views exactly.

This crisis in the life of the Council called for unprecedented action. Steacie wrote directly to Prime Minister Diefenbaker on the situation, in surely one of the bluntest letters ever received by a prime minister:

For forty-three years we have been under the control of an Hororary Advisory Body composed of some of the most distinguished scientists in Canada. A most important function of this advisory body has been the selection and supervision of the Council's personnel. The Heeney report by suggesting the arbitrary transfer of this phase of our activities to the Civil Service Commission without citing any argument as to why this would contribute to our efficiency, has naturally caused both worry and offence to members of our advisory body who have given years of devoted service without compensation.[18]

He told the prime minister that the NRC was a unique Canadian institution; over the years it had built up an enviable reputation, national and international, both for scientific accomplishment and for the efficiency of its operations. In fact, Steacie wrote, on all questions of management of a scientific institution – appointment, promotion, organization, efficiency, and understanding – the Council was far ahead of the commission. He suggested that the commission 'should try to get its own house in order, to try to make itself reasonably efficient, and should try to learn from the experience of other bodies.'

Steacie's main concern in addressing the prime minister was, of course, the Council. He took the opportunity, however, to recommend that other scientific services, such as those performed by the Department of Agriculture, which had done good work in spite of labouring under the control of the Civil Service Commission, should also be exempt. This point – that some scientific structures lay under the dead hand of government control while others did not – was an established fact of institutional life in the 1950s, and a sure candidate for resentment.[19] Such an invidious situation was but one of many which Steacie had to contend with throughout his term of office, but which he was well able, knowingly or unknowingly, to take in his stride.

Meanwhile, for Council administrators it was a time for marshalling of ranks. Rosser sent copies of their memorandum to the other outside agencies, including the Defence Research Board, the National Film Board, and the Fisheries Research Board, asking them to keep the document confidential. F. McKim, NRC assistant director of administration, sent to London for information on the impact of exemption and inclusions of scientific institutions in the British civil service. The replies confirmed exactly the Council's stand. Steacie lunched with Bob Bryce, and gave him a copy of the comments.[20] By the end of February, few people on the scene were unaware of Steacie's personal position on the matter.

In reply to Steacie, Prime Minister Diefenbaker reassured him that the committee of ministers reviewing the Heeney Report had included Gordon Churchill, 'who is familiar with the problem in respect of the NRC ... I can assure you that the views you have expressed will be taken thoroughly into account before any decisions are taken to carry out what is recommended in the Report.'[21] On the same day Churchill also wrote to Steacie informing him that a meeting which was attended by Heeney himself had been held to discuss the problem: 'He [Heeney] retreated from his position and it is unlikely that any action will be taken such as he proposed in his Report.'[22]

Neither the prime minister's nor Churchill's words convinced Steacie that the matter could rest there; the reasons became clear a short time later. What Steacie wanted from the government was not just an assurance for the present crisis – vital as that was – but a promise and a commitment for all time. Steacie wrote again to Churchill:

The immediate danger which we have been fighting is that we should be brought under the Commission. There is, however, a further point. I would be very much upset if any amendment to the Civil Service Act were to contain an enabling clause whereby the Research Council or other agency could be brought under the Act by Order-in-Council. We have, I think, always had a great many friends on all sides of the House and I feel our position is important enough that it should never be changed without a full debate in the House on the affairs of the National Research Council. I would, therefore, like to urge that if any enabling clause is to be in proposed legislation we should be specifically excluded so that it is not possible for the Research Council to be brought under the Commission other than by an amendment to the Research Council Act itself. It is true an assurance from the government that it had no intention of bringing us under the Civil Service Act might be all that is required but it would not guarantee us against irresponsible action by some other Government in the future. I would, therefore, feel very happy if I felt we were protected by actual legislation.[23]

Was Steacie being unreasonable in the light of the minister's generous reassurances? Was there a degree of paranoia in his insistence on pushing for protection by legislation? An exchange of correspondence dating from 1947 between then president of NRC, C.J. Mackenzie, and Heeney, then secretary to the cabinet, gives us an insight into Steacie's thinking and action. On that earlier occasion, Heeney reminded Mackenzie of long-standing plans to reconstitute the Committee of the Privy Council on Scientific and Industrial Research to which NRC was, in theory, responsible but which had not convened during the war years. This would provide, he wrote, 'the Committee with suitable terms of reference and, pehaps even more important, organizing the procedure of Committee through the Cabinet's Secretariat.' An order-in-council had already been drafted and Mackenzie's comments were invited to the points outlined by Heeney, including the suggestion that 'all research proposals be submitted to the Committee ... because it is by such means only that the real job of co-ordination can be done.' The three main agencies concerned were to be the Research Council, the Defence Research Board, and the Atomic Energy Control Board, for which the committee held responsibility for personnel. While routine details need not be referred, the committee would nevertheless concern itself with 'major matters of personnel policy.' The new procedure would also require approval of expenditures in the form of reports from each of the bodies concerned.[24]

Mackenzie's reply was a study in the art of non-commital diplomacy at which he excelled. 'I have no constructive comments or suggestions to make,' he wrote. 'My own view is that by making provision for the formal appointment of a Secretary from your Privy Council office it will be pretty much a matter of how well the Secretary can make the Committee function most effectively and with a minimum of burden on the Minister.'[25] An order-in-council curtailing NRC's freedom of action was obviously not without precedence. Steacie doubtless had this episode in mind when he wrote to Churchill, pressing for firm legislation. From Heeney's point of view, the process of co-ordinating and supervising all matters relating to government, scientific or otherwise, and creating a smooth, orderly machinery of procedure was his objective from the very beginning. From Steacie's point of view, the bureaucrats wanted their pound of flesh at whatever cost to the Council's well-being. If it were to be a battle for survival, it would be an exhaustive one.

At every stage of these developments the Defence Research Board (DRB), a wartime child of the Council, and the closest institution to the NRC, was kept informed. Steacie and the chairman, Hartley Zimmerman, were lifelong friends, a friendship which went back to pre-university days. Not surprisingly, DRB's own minister, George Pearkes, in his well-meaning memorandum some days later to Zimmerman, had brought little comfort, raising the very point on which Steacie had addressed Churchill. 'I am assured that before any change in the present situation is made every opportunity will be given the officers of that group to present their case to the Commission and that no group may have its exemption removed except by Order-in-Council.'[26] Steacie's fears appeared confirmed. Nor did he take any solace from Churchill's reply to his own letter: 'I do not approve of this method in Legislation [of bringing bodies under the act by means of order-in-council] and I think you may rest assured that it will be opposed.'[27]

Meanwhile, communiqués continued to arrive from the network of NRC scientific liaison offices in London and Washington. Harry Williamson from the U.S. office told Rosser of a luncheon hosted by the Commonwealth Scientific Office for Alexander Todd, chairman of the British Advisory Council of Scientific Policy. 'I mentioned what the Heeney Report proposed for NRC. Paltridge (Australia) said that their CSIRO was outside the Civil Service and a bl—y good thing too.' Sir Alexander had remarked that 'DSIR or its predecessor had not always been Civil Service and that it had been a mistake to make it so particularly as the only reason for doing so was to have things tidy.'[28] It all sounded uncomfortably familiar. The reference to the fate of the DSIR was particularly perturbing. It was, after all, the parent, the original idea out of which NRC grew.

Throughout that summer the press kept the issue alive, with the *Montreal Star* and the *Ottawa Journal* each carrying several editorials and a number of lengthy articles all taking the side of the Council. It was a case of 'punching the Civil Service time clock or burning the midnight oil,' as one reporter put it.[29] Ironically, not only the press wholeheartedly supported the Council's stand; of the people they quoted who took the same stand, some were members of the civil service.

In the midst of these controversies, Steacie received encouragement from an unexpected quarter. Dana Wilgress, then head of a government mission researching ways to improve co-operation in international scientific and technical research, and Heeney's long-time friend and colleague, had recently come up with some final conclusions: scientific organizations should be free from government restrictions, should be able to fix their own salaries and hire their own staff. The *Ottawa Journal* reported that Wilgress's recommendations to the authorities included 'just the type of independence for scientific research that Dr. Steacie was demanding.' Wilgress, after a careful examination of the various systems abroad, had concluded, 'I am convinced that we should retain in Canada the very effective and unhampered organizational setup which the Research Council has always had. I heard glowing reports of our Council wherever I went.'[30] He was understandably a little embarrassed to discover that his official findings abroad now appeared to make him partisan to the wrong side in the tussle at home.

The *Ottawa Journal* suggested that Steacie's stand on the matter was so urgent that any change of significance might result in his resignation.

Actually, there is every indication that Steacie was far from contemplating retiring from the debate. If he felt vindicated by the large showing of public and press support, he did not pause there. There can be no doubt that he was genuinely angry. To the suggestion in the report that the commission (by its nature) possessed expert knowledge in the field of administration, Steacie exploded, 'like hell it does.'[31] But Steacie was not averse to a bout of confrontation; he was after all, an Irishman. And there were ways and means.

During August, Steacie gave three talks on CBC radio expounding his views on the role of the NRC in Canada, in which thinly disguised opposition to the Heeney recommendations abounded. His reputation for the abhorrence of red tape and uniformity was by then well established – it merely needed a few embellishments. The opportunity and timing were perfect. Steacie gave an admirable performance, sober yet humorous, eloquent but to the point. It gave vent to his most vehement thoughts on how science was affecting society and vice versa. On the organization of science, he was explicit. 'Probably,' he told his audience 'the most horrible thing that can be done is to bring in a firm of management consultants who know nothing of science and ask them to organize you. The work of the laboratory will be lousy.' His astonishment and contempt for such techniques were genuine, the audience soon discovered. One management specialist, he declared, had actually said in a speech that you didn't have to understand an operation in order to make it efficient. And he added, the only reason why industrial research was reasonably efficient was that attempts to organize it had, providentially, been relatively unsuccessful.[32]

Disputes between the Civil Service Commission and institutions such as the NRC were not new. But an uncharacteristic element in this particular episode was that Steacie clearly directed some of his anger at Heeney personally. For Steacie, it was a departure from the norm. He certainly possessed a fiery temper when angered, but afterward there were never any recriminations or grudges against the offender. The personal assurance he had received from Heeney, which he now took to be an evident untruth, was to poison the present situation. On more than one subsequent occasion when Steacie and Heeney met at social functions, they were observed to walk off in opposite directions. It was a rift which never truly mended until perhaps the last year of Steacie's life. In the mean time, Heeney had returned to Washington for a second term as ambassador and was to suffer a serious illness that same year. Later, when he heard that Steacie was to attend a meeting in Washington, Heeney issued a special invitation to dinner. Steacie wrote to say he was unable to attend.

Heeney's place as chairman of the commission was taken by S.H.S. Hughes. He wrote to Steacie suggesting that further consultations with deputy heads were desirable before proceeding with a revision of the Civil Service Act, since at the time of the original meetings opportunities for study may not have been sufficient to 'produce conclusions which now have occurred to many of the people most closely concerned.'[33] If Steacie felt some umbrage at the NRC being once again addressed as a department of government, he restrained himself from comment. Instead, he suggested it would be more constructive to have a private discussion on the NRC's position. 'I am afraid we have reached a very considerable degree of disagreement with Mr. Heeney regarding his report,' he told Hughes, and added diplomatically, 'I very much welcome the chance of starting discussions afresh with you.' It was pertinent to point out, but gently, the difference between government departments and the outside agencies, as he saw them. The latter were not concerned primarily with the commission's present or future practices; rather it was a case of making clear 'present personnel policies and at least in our case of explaining why we consider that any change would be undesirable.'[34] Hughes, no doubt well aware of Steacie's stand on the matter, was conciliatory, agreeing to a private meeting. 'As I see the recommendations of the Report they are advisory only and I have no indication that the government plans to adopt them all without further close study. On several occasions you made your position clear ... I feel sure [your views] will be sympathetically considered by this Commission.'[35]

During all this time, no comment appears to have come from Heeney himself. He was, no doubt, somewhat surprised by the turn of events. The strongest reaction to his lengthy report appeared to be emanating entirely from a body mentioned only in the appendix. He was, understandably, getting a bit fed up with the press coverage and publicity. But if Steacie had acted uncharacteristically by bringing the controversy into the public arena, Heeney's next action was perhaps even more astonishing. He retorted to Steacie's 'rather sharp views' by sending his side of the story directly to the press. 'We did not say that the NRC ... had no case for separate administrative arrangements. What we did say, in effect, was that, before such exemptions from the general regime were granted or continued, the agencies ... should be required to make their case.'[36] Steacie had, of course, done precisely that, in the most vigorous manner possible, and was prepared to do it again: 'In a major scientific institution the main thing is to develop a character and an atmosphere which distinguishes the organization from all others ... the most undesirable thing in any creative organization is uniformity ... [which] ... should be regarded as an unavoidable evil rather than as a desired goal ... above all I refuse to submit to the view that uniformity is a good thing in itself whether it be administrative or otherwise. If uniformity is the goal of progress the future of mankind looks unbelievably dismal ...'[37]

Heeney was either unaware of these views, which hardly seems likely, or remained unconvinced. 'The NRC case, so far as I am concerned, has still to be made and I am sorry that Dr. Steacie has felt it appropriate or necessary to take the public attitude he has, instead of having his problem looked into calmly and objectively.' Heeney was concerned, perhaps rightly, that the kind of open controversy between Steacie and himself was detracting public interest from the report itself, which was not primarily concerned with the NRC. Irritation was not felt solely on Steacie's part:

I confess that I find something less than convincing as well as ungenerous, [in] the assumption that elsewhere in the Public Service there is no comparable need for the conditions which permit and encourage creative work and imagination. Dr. Steacie should know that it is almost the rule not the exception for heads of government agencies to claim that the special character of their work and their personnel require just the same freedom from central controls that Dr. Steacie demands so eloquently for his scientists ... many of the arguments made by Dr. Steacie are made with similar conviction though perhaps less vehemence on behalf of other elements within the Public Service each with its own particular gloss.[38]

In his reminiscences of the same episode, written over a decade later, Heeney does not mention the controversy involving Steacie, but throws some light on his own philosophy. 'The Commission's staff was imbued with its responsibility as the guardians of the inviolable rule that merit alone should be the criterion of appointment and preferment ... if many of them cherished and revered excessively the intricacies of the regulatory system which had been built up to maintain the principle and prevent abuse, the motive and origins of their zeal and even their rigidity were not only respectable but admirable.' It was hardly surprising that in the process some of the celebrated traits of bureaucracy would develop. During this brief interlude as an administrative reformer, Heeney found himself 'becoming more and more evangelical,' and grew increasingly convinced 'that in the Civil Service we had an institution of great and virtually unknown potential in the continuing effort to build and maintain a united Canada.'[39]

In this, Heeney's and Steacie's final aim may have been the same, but their idea of how Canada was to be strengthened remained very much apart. Normally imperturbable under duress, on this occasion Steacie had felt the situation warranted a reaction beyond calm objectivity. When matters could be viewed at last in this light, time was no longer on Steacie's side. Heeney, a few years later, generously acknowledged his opponent's cause. Steacie's evident 'devotion to our country, his constant burning concern for the public interest' and personal courage would remain high in Heeney's estimation. At the end, 'there was no one in our Ottawa company for whom I had greater respect.'[40] In the final analysis, Steacie won the day. The government did not implement any changes to the NRC. But the victory was neither total nor permanent. One way or another, change hovered on the horizon.

What was to be learnt from this long-drawn-out dispute between Steacie, spokesman for science on one side, and Heeney, the guardian of civil service procedures on the other? With the wisdom of hindsight a number of things are clear. To a large degree, rapid growth and continued success of a complex organization such as the NRC through the 1940s and 1950s had been made possible by compatibility of personalities: initially between C.J. Mackenzie and C.D. Howe; then for a short time between Howe and Steacie; and subsequently by a certain understanding between Steacie and the next minister, Gordon Churchill. That a confluence of shared ideals should shape the history of an institution the size of NRC for two decades is remarkable. It also explains what happens when the supports of such a structure are removed. The separation between science and the state has been recognized since the earliest of times. When united by mutual need, acceptance, and trust, or during threats of war, the situation rests; in times of instability, the thin veils of understanding are found to be inadequate and uneasiness pervades. As an organization, the NRC found itself in just such a situation in the early years of the 1960s.

11
Years of Fulfilment

Being head of an organization such as the National Research Council was by no means a dull round of serious finance planning and high-level committee meetings. On a few occasions, when decorum permitted, even this last unpopular duty could be turned into a lively occasion. Steacie strode along to one such session swinging a bottle of stronger than usual stuff, 'in case the boys got thirsty.' As a rule, however, Steacie guided these meetings from the chair with a resolute hand, refusing to allow the proceedings to give way to discord, the fate of many a wasted day. He rarely objected to repeating the issues at stake or summing up convoluted digressions. These occasions demonstrated his particular gift for clarity of thought and single-minded purpose, unimpeded by side-tracking. Still, even his patience could be exhausted; then the more observant members would see Steacie venting his frustration on his pipe under the table, ramming the tobacco in with grim force.

There was diversity, too, in his range of leadership powers. In January 1958, for example, he wrote at length to the secretary of the Treasury Board. True, he had communicated with him not long ago, regarding an increase in Council's university budget to almost $6 million. But this was a different matter, relating to an NRC technician, the victim of an unavoidable accident. This had ruined certain items of clothing, which Steacie described in detail, listing items destroyed or damaged. The president requested that Council be permitted to reimburse the said technician to the tune of $19.75, which required prior approval by the chairman of the Committee of the Privy Council on Scientific and Industrial Research.[1]

There was also that aspect of every science administrator's job: dealing with the inventions of the public. This task was particularly troubling in wartime. It called for judgment – quickly and reasonably accurately – of the large numbers of claims by sincere individuals that their special idea would help solve problems, if not win the war outright. These 'mad wild schemes' could not be summarily dismissed. After all, both radar and Barnes Wallis's incredible bouncing bombs had evolved from just such unlikely inspirations. For Steacie, there would be no shortage of unsolicited mail from hopeful theorists and inventors. The most persistent of these, one Fernand Roussel of British Columbia, a man without formal education who had developed a new theory of atomic structure, considered it a grave injustice that he had not won the Nobel Prize for his efforts and sought support from Prime Minister St Laurent, who passed it to Steacie for judgment. It was always a painful duty to dampen such unfulfilled ambitions.[2]

Then there were the episodes of pure controversy. Of the diverse changes which marked the post-war era, the advent of atomic and nuclear power stands as a beacon, a scientific and technological enterprise synonymous in Canada with the National Research Council. Steacie was there at the beginning, in the thick of the Canadian involvement. When he expressed the desire at the end of the war to put some distance between himself and the whole undertaking, it was not to prove easy. Scientifically speaking, Steacie, as C.J. Mackenzie pointed out, had never really been interested in any of this 'atomic stuff.' Be that as it may, the subject now would never be entirely divorced from his realm of responsibility. When news of his imminent appointment as vice-president of the NRC was announced in 1950, the press made much of his wartime role in Canada's atomic program. 'Top Atomic Scientist Heads NRC Research,' 'Atomic Expert Gets New Post,' 'Top Atom Scientist to Handle Research on Atomic Fission,' so ran dozens of headings – titles Steacie would never have claimed for himself. In a life heavily filled from minute to minute, Steacie had neither the inclination nor time for abstract ruminations on the more controversial aspects of this science, or rather technology. But in 1950 Steacie had found himself once more quoted in the press on the question of atomic secrets. On this topic of secrecy in science generally, he had plenty to say. Such restrictions in an area of knowledge like science had perverse consequences. Taken to its logical conclusion, as Steacie was always inclined to do, it meant ultimately that American physics would differ from English physics. And nearer home? 'The next step obviously,' he pointed out coldly, 'is for Ontario to refuse to allow Quebec to know the results of work financed by the Government of Ontario, etc.'[3]

In the summer of 1955 Steacie was invited to deliver the opening address at the upcoming conference organized by the Canadian Institute on Public Affairs at Lake Couchiching, north of Toronto. It was to be a week-long meeting dedicated to 'consideration ... of both destructive and constructive possibilities of nuclear energy. In short, the meaning of an atomic age for a world whose peoples do not yet comprehend its full significance.'[4] The talk was to be broadcast by the CBC and in parts of the United States. Steacie intended to give the speech on Saturday morning and lead informal discussions on the peaceful implications of atomic energy on Monday, before leaving at lunch-time to savour the peace and beauty around Georgian Bay.

That the public was in a state of 'complete confusion' over the whole atomic issue was, Steacie conceded, true enough. Personally, he viewed the situation with his usual pragmatism. 'Our dilemma today is that modern technology has placed a vast destructive power in our hands. The atomic bomb is merely the present outstanding example of this.' What did history have to contribute to the dilemma? In the first place, it had to be remembered that inventions appeared as a consequence of a specific need. The development of the steam engine, for example, fuelled the power for mechanization of industry during the Industrial Revolution. An observer, Steacie told his audience, may have noted that the cotton gin was invented at this time, leaving social upheaval in its wake. Surely, Steacie asked, no one would seriously refer to the period since the invention of the cotton gin as the Textile Age? The social problems confronting the twentieth century as a result of technical changes were, in Steacie's eyes, no different in kind from those plaguing society in the centuries before. Such threats as were perceived were actually due to the accelerating pace of change confronting society, and not the result of any specific invention. The point Steacie wanted to emphasize was that this latest technical prowess, nuclear energy, was only one facet of the development of recent technology. 'The essential developments are not the discovery of atomic bombs but the mechanization of warfare and the automation of industry.'[5]

One clue to Steacie's pragmatic stance on nuclear issues was undoubtedly the relative absence of information on radioactive fallout. At the site of the atomic tests, spectators had been provided with heavy eye protection and little else; only much later did the significance of exposure to radiation slowly dawn. By early 1958, however, sufficient universal concern had mounted for a symposium on the phenomenon of fallout to be called by the medical sciences section of the Swiss Academy. The life sciences, endowed now with knowledge of the DNA structure, were facing monumental reappraisals.

'It is interesting to speculate on what would have happened if nuclear energy had not proved to be possible,' Steacie told the audience at the Couchiching conference. 'My personal feeling is that it would have made very little difference. With modern development of electronic techniques it seems certain that guided missiles with conventional explosives would ultimately have had destructive powers not far short of what can now be accomplished by the atomic bomb.' Steacie's opinions in 1955 are, today, deeply perplexing. Steacie did not question that the advent of the atomic bomb raised difficulties, 'The real problem facing society,' he maintained, however, 'is to adjust to the increasing advances and demands of technology, whether in developing weapons or in the civilian aspects of the situation.' That the same society was being asked now to adjust itself to, as he put it, 'man's increasing ability to control and guide the forces of nature' spells a contradiction in terms. Either things are out of control and society is forced to adapt or, if society is guiding and controlling nature, it ought, in all common sense, to be adapted to society and not the other way around. But Steacie remained convinced. 'It seems to me,' he concluded, 'that there is not, and never will be, a period which can justifiably be called the "Atomic Age."' Steacie's personal stoicism is evident, but in this particular instance the difficulty is in integrating his views.

Whatever name that era would go by, society did not, as Steacie might have hoped, share the same calm, detached, unemotional view which he extolled. When the conference ended, Steacie returned to other matters on his busy agenda, innocent of what he was leaving behind. The news media across the country, however, were about to have a field-day. Steacie would soon be astonished to learn that a general remark, made in passing, had aroused a furore. 'Marriage discouraged between atomic workers,' the *Ottawa Citizen* announced. The statement that 'this is a safeguard against the remote possibility that atomic radiation will affect the genes which transmit hereditary characteristics of children' was attributed to Steacie. 'The effect of radiation from atomic bomb blasts is a cause for real concern, the scientist said. The precise nature of such damage is not known.'[6] When the papers hit the news-stands, Chalk River hastily issued denials and corrections. 'Chalk River hands may wed without fear,' the *Ottawa Journal* now contradicted. But too late; papers across the country had latched onto this piece of sensationalism and it was making news from North Bay, Ontario, to St John's, Newfoundland, from Sault Ste Marie to Moose Jaw. A few days later, Steacie's remarks had found their way to Auckland and Wellington in New Zealand, to the *Straits Times* in Singapore, and the *Jerusalem Post* in Israel.[7] *Time* magazine tried hard to reach Steacie for its sensation-seeking audience. When Steacie returned to Ottawa, the worst of the frantic activity by the press was over, apart from a few of the doggedly persistent. 'I made no reference to radiation effects in my address,' Steacie told one American correspondent irritably. 'The statements in the press are a complete misquotation of some remarks made in informal discussion.'[8] Steacie was relieved when the whole silly episode faded.

Neither public hysteria nor private concerns over nuclear issues were new to Steacie. Experimental tests in the early 1950s were frequent and lavish. Activities in the United States had not gone undetected at the NRC. Early in 1955 Steacie had been at pains to reassure the member of Parliament for Churchill, George D. Weaver, who wrote 'regarding the threat to life on the planet inherent in thermonuclear weapons.' Weaver had already raised the matter with Steacie the previous year. Information available then suggested that an explosion of '75 Hydrogen bombs would pollute the atmosphere to a degree that eventually life on this planet would be impossible.' Rumours abounded that Russia would soon 'have a stock pile of some 200 of these devices.' Weaver asked for opinions on the number of bombs required to cause radioactive pollution for approximately six thousand years. This was a matter for the Atomic Energy Control Board, to which Steacie turned. His eventual two-page reply assured Weaver that the production of a worldwide radiation hazard would require uniform distribution of the radioactive debris over the earth's surface. This, Steacie felt, would be extremely difficult if not impossible to arrange, 'unless very large numbers of bombs were used.' The information released by the United States Atomic Energy Commission suggested that a serious worldwide radiation hazard would require the simultaneous detonation of several hundred thousand atomic bombs. 'It therefore appears,' Steacie had ended comfortingly in 1954, 'that there are no immediate grounds for worry on these lines.'[9] His fundamental opinions on the matter had not changed in the interim. However, 'since that time,' he told Weaver, 'it appears that the power of more recent Hydrogen bombs is much higher than was at first thought ... I think it is probably fair to say that the present opinion is that it is most unlikely that even a full scale atomic war using Hydrogen bombs would produce sufficient radioactivity to make life impossible over an extensive area ... I think there is no question that the genetic effects of a wide scale use of Hydrogen bombs might be most unfortunate.'[10]

Steacie held passionate opinions on the responsibility of scientists for their discoveries. The task, and only task, of the scientist was to do good science. What happened to scientific discoveries was up to society. 'It is not possible to develop science piecemeal, or with a specific end in view,' he argued; 'at any given time society has at its disposal a certain body of natural knowledge. It can use this for any desired purpose, military or civilian. Weapons, therefore, are a part of technology, and as technology develops so will weapons. One cannot, for example, have 17th Century weapons without having 17th Century public health.' To call a halt to the scientific development of weapons was, in Steacie's opinion, impossible. As for the 'guilt' of scientists, Steacie was adamantly opposed to any such intimations; a scientist's social responsibility was no greater – or less – than that of any other individual. 'It is no more sensible to criticize the scientist for obtaining fundamental information which may be used for purpose of warfare than it is to criticize a dairy because a bottle of milk is a convenient weapon with which to hit someone on the head.'[11]

Steacie's fatalistic views on the impact of a full-scale atomic war and the impossibility of stopping weapons development were not shared by some of the most eminent scientists of his generation. Close to Steacie's own field of chemical endeavours was Linus Pauling. 'The damage that would be done to the world by a nuclear war is inestimable,' Pauling wrote.[12] By the mid 1950s Steacie and Pauling had been acquainted for some years; they would become personal friends. At about the same time that Steacie was reassuring Weaver, Pauling was not to be so simply comforted. He had already become deeply involved in the issue of scientists' role and responsibility in the problem of nuclear weapons. For his pains, he was branded a left-wing sympathizer, relegated to the class of enemy, and had his passport temporarily withdrawn. The Nobel Prize for chemistry in 1954 had the expected impact. Pauling was elevated from the ranks of the private, but cranky, scientist to those of the public seer; from the obscure to the famous, and by the American government, from the nuisance to the dangerous.

Pauling's growing concern was far from unique. As early as 1945 scientists, including Leo Szilard, James Franck, Glenn Seaborg, and Eugene Rabinowitch, had voiced their deep disquiet to the authorities. By July 1955 the determination of scientists to bring the issue firmly to the attention of the world authorities had manifested itself in a declaration signed by some fifty-two scientific notables, Nobel laureates to the man (and one woman, Irène Joliot-Curie). It was a singularly clear message: 'With pleasure we have devoted our lives to the service of science. It is, we believe, a path to a happier life for people. We see with horror that this very science is giving mankind the means to destroy itself. By total military use of weapons feasible today, the earth can be contaminated with radioactivity to such an extent that whole peoples can be annihilated.'[13] If Steacie knew of this declaration, he remained unmoved. 'There is a great deal of loose thinking on the question of the moral responsibility of science and scientists for things like nuclear weapons,' he told an audience in 1957. 'All science can do is to increase the fund of natural knowledge and thus increase our potential control over our environment. What society does with this power is a social problem.'[14]

In 1957 Pauling initiated the petition which was delivered to Dag Hammerskjöld, then secretary-general of the United Nations, signed by well over ten thousand scientists around the world, suggesting an international moratorium on atomic tests. Steacie, along with other eminent men of science of the day, received a copy, which remains on file. Pauling is almost certain that his friend Steacie had added his signature.[15] If so, it is probable that Steacie put his name forward as an individual scientist, without compromising his position as leader of the National Research Council.

His views, however, had not shifted from those expressed in 1955. In that year, the Pugwash conferences came into existence. Initiated by Bertrand Russell, they were strongly supported by a number of scientists who felt a distinct pang of conscience at having contributed to the nuclear threat now facing the world. It sprang from a desire, however impractical, 'not only to close Pandora's box but also to capture and return to it the unsavory denizens that had escaped when the scientists of the Manhatten Project lifted the lid.'[16] These activities were supported philosophically and financially by the wealthy industrialist Cyrus Eaton, after whose estate in Nova Scotia the conference was named. This association with Cyrus Eaton, whose much-publicized admiration for Nikita Khrushchev lifted even a few liberal eyebrows, would prove unfortunate. Officials pointed significantly to the fact that earlier Pugwash meetings were closed sessions and detailed information on the workings of the conference was difficult to obtain. With expansionist communism rampant in one land, and McCarthyism stalking the other, the 1950s would prove that the cold war could become uncomfortably warm. Opponents labelled the meetings 'Pugwash Fantasies,' convinced that Eaton and his supporters aimed solely to discredit American security agencies, and 'to encourage American scientists to defy them.' An analysis prepared for the Committee on the Judiciary of the U.S. Senate in 1961 reflects with clarity the mood of the day. With names such as Bertrand Russell and Frédéric Joliot-Curie fluttering over the proceedings, Soviet backing appeared self-evident.

Little attention was given to the fact that Pugwash was endorsed by large numbers of eminent scientists, many of whom had no allegiance to left-wing causes; both Jerome Wiesner and George Kistiakowski, close scientific advisers to presidents Eisenhower and Kennedy, attended Pugwash sessions.[17]

The second Pugwash conference was scheduled for 31 March 1958 at Lac Beauport, Quebec. When Lord Russell sent Steacie an invitation, his reply was unequivocal: 'Regret inability to participate since I disapprove strongly of conferences sponsored by Cyrus Eaton, Steacie.' It is possible that the paths of Steacie and Cyrus Eaton had crossed in some professional context. Certainly Eaton was no friend of C.D. Howe's. For his part, Howe ensured that Eaton would not enter 'the charmed inner circle of Canadian business.'[18]

And yet there were elements of these gatherings which would surely have appealed to Steacie. The idea first suggested by a Soviet delegate, that scientists could contribute something which politicians and statesmen could not, 'the honesty that comes from the scientific style of thinking and the independence that comes from the application of the scientific method and the special training associated with the discipline,' would be a recurrent theme in Steacie's personal philosophy.[19] But his views on the nature of atomic weapons and the scientists' role in their fate would be strictly his own. Only a very few of his contemporaries would share this particular brand of pragmatism. Even his former colleague, the indubitably sensible John Cockcroft, would find himself deeply immersed in the whole nuclear issue, and would eventually be elected president of the Pugwash undertaking.[20]

Steacie's stance rested on the firm belief that scientists' business was solely science. On this particular nuclear issue, however, even the greatest of scientific minds felt the pull of their conscience. Einstein had put it thus in a letter sent to leaders in science throughout the world, including C.J. Mackenzie, in 1947: 'Through the release of atomic energy, our generation has brought into the world the most revolutionary force since prehistoric man's discovery of fire ... We scientists recognize our inescapable responsibility to carry to our fellow citizens an understanding of the simple facts of atomic energy and their implications for society.'[21]

Whatever stand he took, all these activities would give Steacie food for thought. In time it became increasingly obvious that the scientist's role could no longer be confined to the laboratory and the pages of scientific journals. This particular strain between science and society, no matter how undesirable, had long ago reached Steacie's domain. 'I must confess that following your letter of January 25th [1954] I felt I had been unduly concerned,' George Weaver told Steacie. But only a year later, 'I am now wondering in view of my responsibility to my constituents, if I was right in placing the confidence that I did in our scientists.'[22]

Steacie, it has been suggested, did not share the guilt of those scientists who had brought the power of the atom into being. Yet as a scientific leader he spoke, not infrequently, for all science. But whatever the ramifications of the anti-nuclear stance of the finest scientific minds of his age, Steacie would remain unmoved; wherever else the growing importance of science and other responsibilities took him, on the question of science, scientists, and nuclear weapons, Steacie remained poised, on the edge of his own personal objectivity.

Steacie's onerous task as president of the NRC did not prevent him from keeping abreast and just ahead of developments in free radical chemistry. One tedious exercise which helped to keep him thus acquainted with his field was the constant request to referee papers. When these did not measure up, Steacie never minced words; if he had graduated to diplomacy in other endeavours, in science it was not the same. 'All in all,' he told editor W.A. Noyes after reviewing a paper carrying an eminent name, 'I think this is about the worst theoretical paper I have seen. It certainly should be rejected.'[23]

Public appearances also kept up his scientific momentum. Steacie was guest speaker of the American Chemical Society in November 1951 at Chicago, and earlier in September had lectured to the Twelfth Congress of the International Union of Pure and Applied Chemistry (IUPAC) in New York. Baker visiting lecturer at Cornell University in 1953, he was named Liversidge lecturer by the Royal Institution of London in 1955. His subject of free radical mechanisms would not necessarily appeal to all, even hardened chemists, but Steacie's style of delivery attracted attention, his clarity of thoughts packaged into precise delivery.

Invitations to lecture increased, as would his circle of listeners. 'The lecture was delightful, being both keenly instructive and most pleasingly presented,' W.A. Hamor of the Mellon Institute commented after a visit from Steacie.[24] The vast authoritative output issuing steadily from his laboratory was the deciding factor, of course, but Steacie's natural ability to perform before an audience, possessed by few scientists, rapidly added to his reputation. 'Were I an agent in the entertainment world I think I would come around and see if I could "sign you up,"' one member of Parliament told Steacie after his television appearance.[25] But this artiste was already too booked up. Being popular meant work: vice-president of IUPAC in 1951, president of the Faraday Society in 1959, and UNESCO and NATO commitments in between.

Steacie's increasing status failed completely to change his lack of pomposity. It was natural to greet the NRC's own janitor and a visiting dignitary in the same friendly way. This simple act is repeatedly told to indicate the degree of its impact; students, fellows, and colleagues would all mentally note and unconsciously emulate. Whatever other memories they took away with them, this particular Steacie characteristic would not be forgotten, nor would his manner of handling awkward situations. Coming into work rather late one morning, one post-doctoral fellow was dismayed to see the energetic Steacie bounding down the front steps, on his way out. Far from showing annoyance on seeing the young man's embarrassment, the president's quick wit greeted him cheerfully, reassuring him with a grin: 'It's quite all right, as long as one of us is here at Council.'

Even on messier occasions Steacie could retain this same imperturbability. This was demonstrated when a water hose came loose one night and flooded the laboratory and the president's newly renovated office immediately below. The distraught fellow prepared for the worst, 'sure there would be hell to pay, for the water damage was a terrible sight to behold.' One wall of Steacie's private library was completely soaked, as were his desk and personal papers, and the carpet was detached from its moorings and in wrinkled disarray. Frantic efforts at salvage were to no avail. Steacie surveyed the scene of destruction with equanimity. 'Relax! these things will soon dry out.' This was not the first flood he had seen in his chemical career. In fact, Steacie took the view that administrative offices should always be situated beneath laboratories. That way, the periodic floods would remind administrators that the real business of the laboratory was science.[26] These little incidents would form a part of Steacie's personal attitude to institutional life, that aspect of it which kept him close to the work he most enjoyed, chemistry.

Abroad, Steacie's accomplishments were increasingly noted. In 1957 his chemical labours were amply rewarded by election as foreign associate of the National Academy of Sciences of the United States. 'I can assure you,' Steacie wrote appreciatively to J.G. Kirkwood, 'that I consider this to be by far the greatest honour which I have ever received.'[27]

On national territory, recognition had naturally come much earlier. Election to president of the Chemical Institute of Canada in 1949 had been followed by the presidency of the Royal Society of Canada in 1954.

To list his honours would become tedious; they were numerous.[28] Steacie was saved from indulging in self-satisfaction by a factor beyond his control: his time was constantly occupied. Communicating with his scientific peers was a desirable and necessary part of a career as chemist, but Steacie had another audience. The public would ensure that the distant ivory tower which Steacie periodically looked at with longing would not be his fate. He could hardly avoid knowing this when there were constant reminders. 'Because of the immense prestige of science in our civilization, the scientist, whether or not he wishes to be, is becoming an important determiner of values,'[29] as one correspondent put it. Personally, Steacie was adamant that this state of affairs was wrong and would protest at every opportunity.

On the whole, Steacie enjoyed the challenge of communicating his subject. How could something like chemical kinetics be explained to a radio audience? The world around us, he saw, served comfortably as a natural chemical laboratory. Chemical changes take place continuously, after all, and everything in the end turns into something else. How fast such changes occur was really the question. The air and gasoline vapour in a car tank, for example, were slowly combining to form the same products as those formed if the gasoline burned. This was not a problem, being a slow process, unless of course you put a lighted match in the tank to see how much gas was left. The high temperature would quickly accelerate the reaction and the speed of change is no longer unimportant, at least to the holder of the match. His listeners usually got the point.

A much more pertinent science-related topic to which Steacie, as president of NRC, gave increasing attention was the organization of scientific knowledge. This was important, both for the layman's understanding of science and to his own philosophy in running the NRC. To start at the beginning, there was the question which frequently worried non-scientific friends: What do scientists actually *do* when they come into the laboratory in the morning? 'Do they sit with their head in their hands until a stroke of inspiration hits them, and go home having made the day's invention?' That, he pointed out to the Wallaceburg Rotary Club, would lead to a lot of head-holding, since the total number of really new ideas coming out of any laboratory was actually quite small. An example of the scientific procedure he offered went thus:

Problem: *To Develop a Mousetrap*
1(a) Look up first 'the Short Introduction to the Trapping of Mice': this most likely would run to over a thousand pages.

(b) Look up information about mice and traps in general, and mousetraps in particular.
II *Pure Research Approach*
(1) Why (or how) do mice get trapped?
(2) What habits are helpful?
Perhaps a paper in some journal – Abhandlungen der deutschen 'mousetrapfen' Gesellschaft.
III *Applied Research*
(1) What is wrong with previous traps?
(2) What can be done to improve them?
These could lead perhaps to a patent or an article in a trade journal such as the official publication of the Canadian Society of Mousetrap Engineers. There was also the short-term applied research approach: could one substitute Canadian cheddar for Swiss cheese?[30]

The audience hugely enjoyed this facetious insight into Steacie's world of science. In its own way, however, this amusing little exercise into the intricacies of scientific research held undertones of greater significance which would increasingly invade Steacie's realm of work. The totally unrealistic and fallacious distinctions between pure and applied research would become less and less rhetorical, as science, society, and politics were thrown closer and closer. Steacie's personal reputation as a scientist would carry the strength of his convictions to the end of his day, but events afterwards would not be so clear cut and certain.

In the fall of 1957 Steacie received notification of an impending honour from an unexpected source – election to the Soviet Academy of Sciences. The timing was interesting. World attention was already focused in that direction with the Russians' successful launching of the ICBM in August. The Department of External Affairs promised to give the news of Steacie's elevation to the highest scientific body in the Soviet Union 'careful consideration.' The Soviets were generally reticent with such invitations to foreign membership, issued few, 'and never in the case of a Canadian.'[31] Steacie meanwhile turned to friends in that most regal and experienced of scientific societies, in London, through NRC's liaison office. How many foreign members did the Royal Society have? Who were they, when were they elected? What was the Royal Society and the British government's general attitude toward membership of a Soviet organization? 'I certainly am all for this kind of opening up of normal relationships,' Steacie told Gordon Malloch of the liaison office, 'but I don't exactly want to be a "test case." '³² The reply from London was encouraging. Malloch had consulted David Martin, secretary of the Royal Society, and its president, Sir Cyril Hinshelwood. In Britain, of its fellows only Lord Adrian was currently thus favoured; Sir Henry Dale had resigned his membership in protest over the Lysenko Affair. Another member, physicist Max Born, had returned to Germany. Steacie would be among the élite of the élites. Certainly, reservation would have prevailed if the honoured scientist occupied a post in atomic energy or defence. Steacie did not fall into either category. External Affairs gave official blessing.

So in October Steacie replied to the Soviet ambassador that the government had approved and he was free to accept the honour. Duly elected in June 1958, friends could hardly desist from pointing to a double accolade which few shared. 'You are a foreign member of the American Academy and now also of the Soviet Academy,' Otto Maass wrote. 'There is no one I know who is more deserving than you are of being recognized by both the American and Soviet Academies.'³³ From the Russian point of view, an exchange of scientific delegations would be the first item on the agenda: Steacie would lead a Canadian group to the Soviet Union in the winter of 1959.

November in two of the coldest capitals of the world offered much in common. The Soviet scientists themselves represented no real surprise either, although Peter Kapitza's marked Scottish accent was unexpected. When it came to the work of academicians such as Kapitza and Nicolai Semenov, scientist spoke to scientist and it was all familiar territory. Steacie was no doubt enthusiastic to confer with Semenov, who had shared the Nobel Prize in 1956 with Sir Cyril Hinshelwood for contributions to chemical kinetics. The quality of work under Semenov at the Institute of Chemical Physics, set in the Lenin Hills overlooking Moscow, certainly lived up to expectations.

The reputation which had preceded Steacie to Moscow was not only in chemical kinetics. The previous year, as president of the Faraday Society, Steacie had requested that a particularly bright young physical chemist in Semenov's Institute be allowed to give a series of lectures in England under the auspices of the society. The Soviet Academy had replied that the chemist 'was too busy.' Now that he was in Moscow and at the Presidium of the Academy, Steacie repeated the request. Officials countered with the name of a substitute, as that other chemist was still 'exceedingly busy.' Well, in that case, Steacie replied, the society would wait.³⁴

Generally, however, the Canadian delegation noted that spirits among

Soviet scientists were good. Stalin's death had reversed recent plunges into disfavour and the scientists had retained their humour. Asked where they obtained the steel for the large cyclotron at Dubna on the Volga River, one hundred and twenty kilometres northeast of Moscow, they replied, 'We melted down the iron curtain.' Dot, who had accompanied the delegation, was meanwhile savouring the cultural side of Soviet life, with visits to ballet schools, old peoples' homes, and the palace where Rasputin had been murdered. Conversations at dinner in Peter and Anna Kapitza's home were enlightening. The influence of Kapitza's earlier Cambridge days was a permanent one, for as often as not the guests dined at a table graced by fine sherry, claret, and cognac, with no Russian wines in sight. The strong-willed Kapitza would have noted some shared characteristics with this leader of Canadian science – an impatience with bureaucracy for example. Steacie, for his part, would have listened sympathetically to Kapitza's methods in dealing with laboratory shortcomings. On returning to Moscow in 1934 and forbidden to return to Cambridge, Kapitza had made his presence at the laboratory felt. Doors which did not fit had been chopped down and concrete floors not level had been hacked up.[35] But no doubt, as with Steacie, time had a mollifying influence. The ostensible purpose of the visit – an accord for exchange of scientists – was achieved before leaving and signed 'with a flourish' on Catherine the Great's golden desk, in the presence of the usual press and photographers. There was time enough to be impressed by a farewell reception and banquet in the splendid oval-shaped white marble room of the Presidium.[36]

It was a constructive time to visit the Soviet Union, whose program of science and technology would have increasing impact on the West. There were also the more positive aspects. 'Russian life,' Steacie noted, was not really 'so different.' There were some distinct advantages – a general lack of advertising, for example. However, it was undeniable that in the Soviet system everybody was a civil servant. He wondered if Heeney 'would be happy.'[37]

But Steacie's view on the fortunes of science in the process of history had altered perceptibly. In the earlier days, his irritation with the hidden and secret domains of science was totally unrestrained. Parasites, he had scornfully labelled those scientists who perused the open literature but did not publish their own findings. Now, however objectionable, he reluctantly saw it was inevitable within certain contexts; economic, political, and particularly military interests decreed that science could no longer remain totally unfettered. It was a source of lasting and genuine regret to Steacie that science, formerly a purely intellectual pursuit, was now endowed with unwelcome addenda. 'Secrecy means nationalism of the most extreme form.' Worse still, the utility of science in recent years had brought it into the public and political domain, previously unknown. Science had become a 'spectators' sport with the box office and publicity looming large.' It would be cited 'as responsible for everything from toothpaste to atomic power.'[38] Steacie's mood of gentle flippancy would soon grow more sober and urgent. A year after his journey to the Soviet Union he was advocating the surprising view that 'in international affairs it may be often necessary to ignore competent scientific advice.'[39] But the danger of weakening traditional scientific structures by excessive narrow nationalism, which Steacie saw so clearly, was no mere bias. This aspect of science and government would grow, as Steacie predicted, with increasing implications for both science and government.

Despite murmurings from near and far during the early part of 1958, Steacie's world remained largely unperturbed. A major issue of the day, and for some time to come in many countries, remained that of manpower shortage in the sciences and engineering. For scientific leaders in Steacie's position, solutions hinged solely on whether universities had the capacity to absorb the impending increase in university enrolment, and how to achieve this without compromising standards or the traditional sanctity of the learning process. There was really no argument on the matter.

Scientists themselves, however, began to experience a new type of pressure. A few notables expressed vociferously their unease at the increasing tendency to treat science and the scientist as commodities, 'with all the appropriate export and import regulations which relate to important strategic materials. The great drive now going on to increase the number of scientists and engineers takes on the appearance of stock piling of tungsten or copper.' Movements in another sphere were growing inexorably in the newly developed art of administration and 'management of scientists.' Protagonists of this form of control were already looking to improvement of their methods. Eventually, this new discipline would generate an approach from the opposite direction. Not merely how to manage scientists themselves, but how economic assessments of science and technology in the national fabric could guide appropriate distribution of scientific effort. Such an excursion into rationality on the part of policy-makers was intended to ensure that 'the disposal of a valuable resource' did not meet the dreadful fate of 'random and undirected influences.'⁴⁰ This impending conflict would lead, slowly and inevitably, to an uneasy and tenuous equilibrium between holders of the traditional views of science and the subject's new master, the government.

Steacie was reminded of the global nature of the manpower shortage problem by a request in January 1962 from the Royal Society of London. He was invited to be the sole non-British participant in a painful British exercise – an ad hoc committee set up to study the emigration of British scientists. Actually, the United States was by far the largest recipient of Britain's intellectual resources in those years.⁴¹ This perplexing issue of the brain drain would continue to raise controversy and high emotion in the British Parliament, in the press, in public and in private, aptly summed up by one wit thus:

A White Paper on our scientists
Is published to explain
What's made them leave old England's shores
When they've worked so hard to train.

Some say they wish to travel
Experience to gain;
Others thought of opportunity,
Position, wealth and fame.

So they left behind their favourite pubs,
Their English country and lane,
Their cricket, beer, and fish and chips
And constant English rain.

But did it need a Working Party
To make these facts so plain?
More cash, more scope, more everything
And in exchange — their brain.

But alas, I'm not a scientist,
And it's driving me insane;
There's all that money over there
And I've no brain to drain!⁴²

Steacie himself had always proved a sympathetic and generous supporter of scientists who wished to relocate. Post-doctorate fellows of calibre were given encouragement and the opportunity to take up suitable posts anywhere they chose. American industries proved zealous in their recruiting methods; British teams sent over from the United Kingdom to entice fellows back were more restrained. A number who returned to Britain during those glorious days of boundless scientific expansion on the North American continent most frequently cited patriotic reasons. As was to be expected, those returning after the NRC experience performed extraordinarily well. Those halcyon days of creative work, shared purpose, and inexplicable good spirits would last them a lifetime.

Steacie replied to the Royal Society a few days later. Distance would preclude his frequent presence at the proposed committee meetings but, he told the secretary, David Martin, he expected to be in England with Dot early in April en route to the International Council of Scientific Unions' Bureau meeting in Rome, and would discuss the matter further then.

There is no doubt that Steacie's perception of solutions to the problem of industrial research in Canada had altered toward the end of his life, and debatably much earlier. At the start, his view on the matter was simple: 'The proper solution, of course, is for Canadian industry to run a small research laboratory and exchange results with its U.S. counterparts,' an ideal which demanded commitment from industry itself.[43] This would shortly be replaced by a more realistic perspective. Even so, his impatience showed through painfully to those sensitive to criticisms. Part of the country's ills could be traced to the notion that 'we are always boasting about our natural resources, for which we can claim no credit,' he sternly told a conference gathered at Queen's University. 'We rarely take the attitude that we should be able to do some things better than anyone else. In some ways the magnitude of our resources has diminished our initiative. This attitude is, of course, a natural historical inheritance, but it is time that it was altered.'[44]

From the mid-1950s on there would be few public occasions when the subject of industrial research would not enter Steacie's speeches. That meeting, with members of the Canadian Council for Economic Studies in 1955, was perhaps critical in focusing Steacie's personal convictions. Although the format of the subsequent memorandum from the meeting makes it difficult to interpret each statement in its proper context, there were a few points of view with which Steacie would never reconcile himself. No scientist worth his salt could help industrial research – or any research – merely with the aid of a sharp pencil, as W. Goforth had suggested before the meeting; and Canada should categorically not be short-sighted and depend solely, or even primarily, upon long-term licence agreements with others. The only level which really counts in science is the working level, as he frequently pointed out. But with wisdom there had come, inevitably, acceptance of undeniable facts of modern life. Science, by becoming important, had also acquired controversial characteristics along the way. As chief spokesman for the NRC, Steacie had learnt, to his surprise, that a basically practical topic such as industrial research could arouse strong reactions. At the Ottawa Board of Trade annual dinner in the spring of 1961, Steacie ventured again on the subject of scientific and industrial research, 'as unemotionally as possible.' This was not easy; as was invariably the case when dealing with economics, any suggestions made were liable to be quoted without qualifying clauses. 'Thus the simple statement that it would be nice to see more industrial research in Canada is apt to be regarded as critical of industry, as ultra-nationalistic, as anti-American and as showing ignorance of economics.'[45]

Earlier Steacie had considered that it was neither possible nor desirable to expand Canadian industrial research too fast. What was needed was a continuing, steady, healthy growth with emphasis on quality: 'Certainly bad research is worse than none.'[46] By the beginning of the sixties, however, the continuing sluggish pace of industrial development was taking on some urgency: 'If we are to have any effect of the situation in 1965 or 1970, we must start now,' he urged. The problems to be overcome were no less awkward than those he had listened to in 1955 and, if anything, they had grown. Steacie remained undaunted by the prospect. 'No matter what the difficulties may be, the present situation is intolerable ... we can't do anything sudden or dramatic, but perhaps we can do something. At least we should give all possible encouragement to industrial research in Canada and perhaps even put on a little pressure.'[47]

One possibility advocated by Steacie offered, if not a wholly ideal solution, at least a reasonable one. It was related specifically to a phenomenon pertinent to the times. In the early 1960s research laboratories in the United States had grown to such an extent that research activities were being decentralized to branches and subsidiaries within the United States. Was there any reason why this decentralization, in certain fields at least, should not be channelled to the Canadian subsidiary rather than, say, the Oklahoma branch? Could not the effort in one field rest with a Canadian subsidiary, while still dependent on the parent in other fields? Certainly a few of the advantages were self-evident. Institutions remained creative and productive only if they did not extend beyond a certain size. This principle had always been behind the sloughing off from the NRC of activities such as atomic energy and defence research.

The other solution had also occupied Steacie's waking hours for some time now. Technically speaking, Steacie saw a direct analogy between industry's research position in 1960 and that of universities in the 1930s. Then, Canada had been faced with students finding their way to training abroad. By much effort on the part of the Council and, eventually, a co-operative government, the situation had undergone a radical change. Universities now stood firm among the ranks, a fulfilment of Tory's own hopes and philosophy. This pleasing chapter of NRC history had largely resulted from one simple act – the provision of financial encouragement. Would this same formula prosper, if transferred to an industrial setting? Much of the problem in Canadian industries lay, as he had repeatedly told audiences, in the branch-plant mentality. True, Canada had been much later in starting, and the proximity to the United States was a major factor. While expenditures on research had increased recently, other impediments appeared immovable. Research was the prerogative of the parent organization outside the country, a fact now considered by Steacie to be by far the most important factor in assessing the position of Canadian industrial research. It seemed unlikely that further tax concessions or exhortation would appreciably increase spending by private industry on research. Perhaps the only way in which the situation could be rectified was by some form of direct government financial aid.

In the fall of 1961 Steacie sent a memorandum to Treasury Board and the cabinet containing his thoughts on federal support of industrial research in Canada. The plans had been some time in preparation. Steacie had sought the opinion of other government agencies and senior executives in Canadian industry, and simultaneously mulled over the implications with old and trusted colleagues such as C.J. Mackenzie, Hartley Zimmerman (now chairman of the Defence Research Board), and J. Lorne Gray, president of AECL. When the analysis of facts and figures relating to research expenditures, both within the country and abroad, was complete, the government was confronted by the fact that financing of industrial research in Canada showed an under-commitment of between ten and twenty times that of the United States and Britain respectively. In GNP terms, the factor was reduced to around three or four. The reasons for this enormous discrepancy lay largely in the inordinately high expenditures by both United States and British governments on weapons research, an outgrowth of both the Second World War and the cold war. Particularly in the United States, government contracts for the military now dominated laboratories such as the National Bureau of Standards in Washington. By 1951, three-quarters of their total effort had been channelled into meeting the requirements of the defence program.

Canada was not in the same position. Instead, Steacie proposed that the government assume, as a long-range objective, the matching of industrial research funding up to a limit of $50 or $100 million a year, to be reached within five to ten years. To avoid wrangling and lost time over monitoring the performance of industries receiving funds, it appeared sensible to Steacie to avoid driving a hard bargain, and thus risk estranging the very body being helped. Essentially, the government funding, if approved, would finance salaries and wages for these expanded programs, but industry itself would pay for all equipment and overheads, thus approximately equalizing the distribution of cost. For Steacie, this would be a fundamental condition of the plan: 'No permanent impression will be made on industry unless it is spending its own money on research.' The plan would begin modestly, with no more than $2 million, since the essential component of the entire undertaking was its long-term nature and the promise of continuity. 'Industry must be assured that provided the work done is good, aid will be forthcoming over a period of from five to ten years, since a team cannot be built up in an atmosphere of uncertainty.'

The idea behind this industrial research assistance program (IRAP) was not fundamentally new. In 1959 the Defence Research Board had hoped to soften the blow to Canadian firms after the cancellation of the huge Avro Arrow program by offering financial assistance. What Steacie was advocating now was an across-the-board measure aimed, as he clearly indicated, at permanently strengthening the country's industrial foundation, and human competence. 'The aim must be to build up real and permanent applied research and development competence ... the whole object of this scheme is to increase the number of research workers in Canada.'[48] When eventually put to the Advisory Panel on Scientific Policy, the final decision was to ask Treasury Board for a sum of $10 million. In the event, the program was formally approved in November 1961, with an authorized sum of $1 million for the first year.[49]

One indication of industry's increasing interest in research was the brief submitted by the Canadian Manufacturers' Association (CMA) to the prime minister in December 1961. A copy was also sent to Steacie for his comments. Among their recommendations, intended to stimulate industrial research and development, was a change in the membership of NRC's council 'to bring its composition closer into line with its statutory title and terms of reference.'[50] In commenting on the brief, Steacie reminded the

CMA's J.C. Whitelaw that NRC had been doing its utmost for almost thirty years and particularly since the war, to interest Canadian manufacturers in research with little success. On the suggestion of increased industrial representation on council, changes were desirable and had already begun. Other requirements, including regional distribution, would mean that further changes would be made gradually but ultimately industrial representatives should be increased to four or five, as advocated by the CMA.[51] That summer Whitelaw invited Steacie 'as Canada's outstanding authority' to deliver the introductory address at the CMA meeting devoted to research for survival.[52]

Of the two solutions to the deficiencies in industrial research proposed by Steacie in 1961, the first, to exhort American firms to settle part of their research establishments with Canadian subsidiaries, was beyond Steacie, NRC, or even the Canadian government's jurisdiction. It was the second, IRAP, that would prove to be the bird in hand. Steacie had taken his model from the Council's long experience in making grants-in-aid to universities. This, begun as the earliest and most urgent of Council's tasks, had in forty-five years (up to 1961) demonstrated unimpeachable success. By the same token, there was now every hope that industry would respond similarly.

Both solutions, however, depended intrinsically upon one overriding factor: the availability of suitably trained personnel. It was this limiting factor which dominated all prospective scientific and industrial development and which, brooking no arguments, had provided the impetus for Steacie's policies and plans of action for a decade. It is of significance, therefore, particularly in the light of the spate of literature on science, government, and policies in recent years, to find that the murmurings of discontent agitating for change had started to be heard in various parts of the NRC domain before 1963, the year of the Glassco Report. Astonishing, indeed, that the nucleus of perturbations would be seeded in the traditionally apolitical world of high academia, the very institution to whose continued well-being Steacie had expended so much energy.

12

Final Days and Unfinished Business

In the fall of 1958, for almost the first time in his life, Steacie had succumbed to illness, which had interrupted his accustomed stride of relentless activity. It had been this great vigour and health, coupled with tremendous self-discipline and a flair for organization, that had enabled him to achieve so much. Now the malignancy of what would become cancer appeared as a growing presence. Early in 1961 Steacie learned that the cancer had recurred, and he would have to undergo major surgery at Johns Hopkins Hospital.

It was an inpropitious time to be absent from Council affairs. The IRAP program, to which he had given so much of his recent time and effort, was due to be put before the Advisory Panel on 19 May. The only comfort was knowing that, having left the whole matter in C.J. Mackenzie's hands, there would be the most experienced person possible to see it through. On his return to Ottawa, Steacie was elated to find that his program had received government blessing, albeit a qualified one. It was unquestionably one of the most encouraging developments of the past year.

Reviewing events for the president's report that year, Steacie reflected on the dual objectives which had dominated NRC efforts and his own: industrial research, and support for university research, which to his mind were inextricably linked. NRC support for university research in the coming year was expected to reach $14.4 million, a fourfold increase over that already spectacular year of Sputnik, 1957 – an improbable rate of progress compared to the $14,000 of 1917. As for Council's own laboratories, 'by far the largest and most diversified industrial research complex in Canada,' they would continue to provide assistance to industry, now poised to building up its own research facilities. But, he warned, industrial research and development would flourish only in a favourable environment. This followed from personal experience. 'The major accomplishments of the National Research Council over the past 45 years have been the fostering in Canada of first rate academic research and the creation of supporting Government laboratories ... the Council has been a major force in creating in Canada the scientific climate and the technical outlook that have carried the country into the research age as an advanced nation.'[1] It would be his last report as president.

The insidious illness which would end his life was accompanied by unexpected side-effects. With will-power, by ignoring pain, it was quite feasible to carry on exactly as before. In between drastic bouts of surgery and subsequent convalescence, Steacie appeared, if not in excellent health, at least to be on the mend. The last attack had been the most devastating; he was incapacitated for almost five months. When he returned in September 1961, it was to a desk laden with work and invitations. That month, Steacie was formally elected president of the International Council of Scientific Unions (ICSU), but pressure of work at home forced him to cancel his planned trip to the upcoming meeting in Delhi. He was confident, however, that he would meet future ICSU obligations. Late in September he flew to London; at the end of October he was in Paris to attend OECD's ad hoc meeting on science policy. There were also elaborate travel plans far into 1962.

The recurrent operations had left him weak, with a great deal of physical discomfort. Steacie pushed on with Council business and made the journey each day to Sussex Drive, though there was now little hope of walking the pleasant mile or so through the woods of Rockcliffe Park. This was a wrench. He had always been active and flourished with a lot of exercise. These morning and evening walks had become one of the most treasured parts of the day, providing time for reflection. One of these evening strolls had almost ended in disaster. An assailant, creeping out of the falling dusk, had dealt the surprised Steacie a mighty knock on the head. The man got nothing for his pains, except a return blow which Steacie managed to deliver. But by the time he reached home, Steacie was bleeding profusely, much to Dot's horror. A few stitches ended the affair, apart from an irretrievably damaged hat. The press reported this singular incident indignantly and Prime Minister St Laurent expressed concern for the welfare of his chief scientist. Steacie took it all with good humour; he was heard debating whether the attacker might have been a disgruntled academic who had not had his grant renewed.

In the spring of 1962 medical opinions were still no more certain, but diagnosis predicted at least another six months, perhaps a year, of life.

Steacie pressed on with work. International scientific affairs had become increasingly demanding – an inevitable consequence of the rising stature of science itself. Steacie had long since recognized that in Canada, a new country separated from the traditional centres of learning, the NRC had a specific role to play in ending the country's isolation. The awareness of the growing status of science, sharpened by repeated visits abroad and the presence of an international band of post-doctorate fellows, had increased with the years. By the beginning of January 1962, just months after his election to its presidency, Steacie had begun the long process of critical reorganization of the embattled ICSU, which was suffering the simultaneous, paradoxical conflicts of both centralization and fragmentation. This international body, purportedly devoted to the sanctity of science, was a long way from giving up national sovereignty even for the sake of avoiding war, as W.A. Noyes would cryptically note. Steacie's plans and recommendations struck, as always, at the heart of the matter. 'The essential thing about ICSU and the Unions is the avoidance of national politics and of a governmental structure. The strength of ICSU is the fact that the Unions automatically provide scientific competence.' The organization of international bodies like ICSU was, to Steacie, no different in principle from that of, say, the NRC. Above all, the executive board should not become so large as to prevent efficient conduct of business. The secretariat needed to be strengthened but 'no grandiose secretarial structure should be contemplated.'[2]

Steacie was uncustomarily depressed by the problems besetting this world body representing diverse scientific interests, even as suggestions from representative nations across the globe arrived on his desk at Sussex Drive. Astonishingly (for all member nations had first to be consulted), Steacie managed to gain accord for his views by May, when he chaired the first Future Structure of ICSU meeting in Paris. It was surely a diplomatic triumph, a masterly show of chairmanship, no doubt drawing on his experiences with the international band of prima donnas of long-ago Montreal laboratory days. Even men of irrevocably opposed political views were apparently united by his 'presence, counsel and wisdom,' as Detlev Bronk, the president of the National Academy of Sciences in Washington, put it.[3]

The ramifications of international scientific bodies were not new to Steacie; he had had a taste of their attending peculiarities by being the first Canadian representative on the executive committee of the division of physical chemistry of the International Union of Pure and Applied Chemistry (IUPAC) and had been elected the first president of that division. For one who eschewed administration, his success in this venture held prospects of uncertain blessing. UNESCO affairs soon followed. Then there would be the scientific committee of NATO, where Steacie's gift for strong leadership would be felt, short as his presence was. From this latter committee, on which Steacie would contribute 'brilliant and imaginative work,' he was now forced to retreat, as health dictated.[4]

The demands of travel drained his resources. On 30 January he had flown to Paris to attend the OECD meeting; now in May he was back for ICSU. In between, his timetable resumed some of its former character – not unlike that of a dentist or doctor, he once commented laughingly to Dot, with an endless series of half-hour or longer appointments filling the day. In June he travelled to Montreal to address the Canadian Manufacturers' Association, felt well enough to throw a cocktail party, and attended a state reception in honour of the Queen Mother.

In July his attention remained partially focused on international affairs, but business at home continued to pile up. To a request for opinions on prospective principals for McGill he gave serious and careful analysis; to recommending and ranking of possible chemists for York University he devoted equal energy. He planned to travel to Prague in September and on to Warsaw on ICSU business; summoned up persuasive powers to gain more elections to London's Royal Society for Canadian scientists; encouraged job seekers – 'we are doing all we can to stimulate research in industry, and I hope this will have a favourable effect on industrial openings.'[5] On no occasion did he indicate awareness that mortality was closing in; his job was clearly not finished, the end could not be.

But very soon Steacie was resigned to responding only to the most urgent business. Even an invitation from the National Bureau of Standards to be principal speaker at a free radical symposium had to be declined. This probably rankled as much as anything could. In his own mind, Steacie remained intrinsically a chemist by inclination well into his presidency and saw his field of work as 'photochemistry.' Now he had to admit that the heavy load of administration had taken its toll. 'The difficulty,' he wrote to friend and fellow chemist Fred Dainton in August,

is ... that you can't go on forever on a 10-hour day doing everything at once. I think I managed to play both games together til I was about 54 or so, and since then I have, of course, quietly folded up as a chemist. My illness had something to do with speeding up the process. However, for one reason or another a decision has to be made. In my case I had little choice, and I must confess that the running of NRC has been fun. Off-hand I would say that I would have been happier if I had remained a chemist.[6]

There had been compensations. Steacie encouraged younger scientists with deeds and words at every opportunity, writing frequent and glowing testimonials. He persuaded one such aspirant, who was feeling depressed, to consider things carefully: 'I am still enough of a chemist to regret sincerely your feeling that you ought to get away from research.'[7] But there was little time to think about research now, although in July he wrote off to Melbourne for chemistry reprints. Things at the Council took up all the energy he had left.

Steacie's initial elation at Treasury's approval of IRAP dissipated as it became clear that the funding was not the precursor to bursts of activity as he had expected. The new plan to aid industry had been mentioned briefly in the throne speech, but otherwise little noticed. His own speech at the opening of Imperial Oil's new laboratories in Sarnia in January 1962, however, had been carried across the country by wire services, and the *Financial Post* had also given good coverage. The staff waited for calls from industry. To their surprise, nothing materialized. But in March, an item which had originated with an American public relations firm informed readers:

a huge saving on research costs will soon be possible in Canada. Beginning April 1st, Canadian subsidiaries of U.S. companies can get grants from the Canadian Government covering the salaries of all scientific personnel working on basic research ... aim is to get U.S. firms to shift projects to Canada.

Over the next ten days, the NRC was swamped with over 250 inquiries from the United States, including federal and state government departments, universities, consultants, lawyers, accountants, even libraries. Things were picking up. Canadian subsidiaries, alerted by their parent companies, were following suit with increasing interest.[8] The IRAP was finally in business.

That summer a new government austerity program hovered on the horizon. Steacie responded rapidly, reducing NRC's budget by $1.5 million in a few days, achieved by trimming administration, construction, and maintenance. The Council was anxious to co-operate in every way possible in the emergency. Treasury Board's decision that future NRC appointments and projects be first referred to the board was, however, a different matter. Financial stringency, Steacie told the secretary of

Treasury Board, did not make Council less competent nor the Treasury 'more competent to judge the relative merits of research projects.' The best solution, Steacie thought, was for Treasury Board to inform the Council of the reduction required in the manpower ceiling at any time and then the Council would undertake to limit its activities to the ceiling assigned.[9] Steacie would never tire of reminding the government that the Council, to do its job, had to maintain independence and internal integrity. But his relationship with Treasury Board remained largely cordial to the end, perhaps because, as Leo Marion suggested, even bureaucrats could appreciate a man who laid things on the line.

The late summer always a good time to be at the cottage, with the president officially on holiday for three whole weeks, promised some of the happiest times. The previous year Steacie had been recuperating and unable to attend meetings with a gathering of visiting scientists, who journeyed to the cottage instead. It turned into a crowded day of sailing, fishing, walking, swimming, and much conversation in which the convalescent Steacie was an unexpectedly lively participant. This August it was hard to recapture that spirit, which would linger as a cherished memory for his guests.[10] Steacie settled down to the solemn task of writing the formal Royal Society obituary of Otto Maass, who had died in July. 'Certainly no one owed him more than I did,' he wrote Carol Maass. 'I feel that anything I may have accomplished was due to his influence.'[11] But it proved impossible to complete the job. By the middle of August ominous symptoms, now all too familiar, intruded persistently. He drove back to Ottawa for the doctor to confirm his own diagnosis.

Steacie took the fact of his impending death with characteristic discipline and the same pragmatic frame of mind with which he had approached all events of life. Back at the cottage, he brought in the outboard motor from the boat and locked it away, after one more deliberate trip around the lake. When everything was in order, Steacie drove back home to Rockcliffe with Dot. At ten o'clock on the morning of Monday, 20 August, C.J. Mackenzie and Steacie met in conference on an issue of overwhelming significance: a successor for the NRC leadership. Both men were agreed that Harry Thode, the president of McMaster University, was by far the most desirable candidate. But the process would not prove straightforward and time was not on their side. It was the last occasion that Steacie and C.J. would see each other. Steacie dictated one more letter to the secretary general of ICSU. 'Unfortunately I have been ordered by my doctor to take a rest for an indefinite period.' Steacie delegated presidential powers to General Laclavère.[12] But at the weekend

Steacie had a visitor from France, General Laclavère himself. Steacie's previous communication of 2 August had certainly not prepared Laclavère for what he saw. Nothing had hinted at the gravity of the illness. It would be Steacie's last public appearance, a 'visite tragique,' remembered vividly by Laclavère over two decades later.

Il était habillé et s'est levé de sa chaise pour me saluer mais il n'a pas pu faire un pas. Il fumait une cigarette. Il m'a simplement dit ces mots, 'I am exhausted and I am dying.' Après cela il m'a donné ses dernières volontés en ce qui concernait l'avenir de l'ICSU puis il a demandé à Mrs. Steacie de nous faire du café. Je suis resté auprès de lui une demiheure à peu près. Il était épuisé.[13]

There was an executive meeting scheduled for Tuesday, 28 August with Gordon Churchill to discuss American and Canadian co-operation in research, and Steacie also expected to see Harry Thode. It was not to be. Sometime during that morning Steacie died in his sleep.

The end, when it came, had been unbelievably quick. Only a few days earlier he had still possessed enough strength to attend to affairs at the office; not many days before that he had attended a seminar given by chemist Bob Cvetanovic and showed his usual interest with penetrating questions. His appearance was that of a tired man, but that was expected. Less than a handful of people guessed these would be final days. When the news of his death spread through the Council, the initial reaction was disbelief. 'It was the biggest shock of my life,' Herzberg recalls. The suddenness of the event left those who knew him numb, from his close colleagues to technicians and janitors, all of whom knew their president as Ned. Herzberg was not alone in the feeling of personal loss at Steacie's death. Those had been happy, fruitful years for Herzberg at the NRC. He was mindful of the wonderfully creative spirit pervading the laboratories, which was, he believed, due to 'Steacie's great gifts and his personal charm ... there was nothing that any one of us would not do for Steacie. He was loved by all; he had no enemies.'[14]

The public, if they were not aware of Steacie's presence before, soon shared with the press the loss which had hit the National Research Council; it was also a national loss. More than one contemporary figure reflected that Canada had lost one of its great men. 'Steacie was one of the most brilliant and noblest men of our time,' Richard Gwyn wrote for the *Canadian Commentator*. The *New York Times* rapidly published a special account of Steacie's career and achievements. Abroad, *The Times* of

London carried Sir Eric Rideal's lengthy eulogy on Steacie's profound influence and contributions to the development of Canadian science, both academic and industrial. But, as with others who had known Steacie and Dot, it was the always warm and generous characteristics which surrounded their personal life that Rideal mourned.

Back in Ottawa and the NRC, it took longer to recover. To the scientific institution which he had led for a decade, Steacie's decisive yet compassionate personality had become the stuff on which other men's dreams were made and hopes fulfilled. He had been both selfless and fearless in providing for their needs, entrusting the staff with his full confidence, supplying the strength of leadership and a willingness to take full responsibility for the consequences. Not only a superb administrator and scientist of high repute, but one without pride or arrogance, rare and precious qualities.

In Rockcliffe, the man who had early on recognized these very qualities in his younger colleague retired to his study and sketched his tributes. 'Ned Steacie was endowed generously with the elements of greatness, his mind was clear and always discriminatory, his thoughts original and he never compromised his opinions.' No one knew better than C.J. Mackenzie how these qualities had succeeded in propelling the NRC forward. Prestige, that nebulous quality, the 'coin of the realm' in internationalism, had made Steacie, as president of ICSU, the world's most distinguished international scientific administrator. But Steacie was, in the true sense, his own man. 'He neither sought nor valued public esteem of an official kind but cherished the friendship of those whose opinion he valued.' It was his friend's human qualities, his wit and charm, 'his sympathy and his unselfish kindness for those in trouble' that C.J. now counted as the greatest loss. Steacie's era was at an end but his influence on the scientific and public institutions of Canada, Mackenzie predicted, 'will not die with him but live on forever.'[15]

In the president's office, it was a painful business catching up on office mail. There were invitations to address the Empire Club, to attend the Photochemistry Symposium at Rochester in March, a returned galley of a book which needed a foreword from Steacie, a letter from Moscow concerning ICSU, and an invitation to a guest lectureship at the University of California. In time, things started to move again, but there were so many reminders. It was difficult to get 'unstunned' as long-time secretary Ethel Wheaton put it. She took on the job of completing unfinished tasks; Steacie was in the tedious process of getting signatures for the Certificate of Fellowship for the Royal Society of London for a deserving Canadian.

She was now so familiar with office procedures that it was almost an automatic response.[16]

Institutions throughout the country hurried to do him honour. McGill, his alma mater, fittingly one of the first, requested permission to name a point on Axel Heiberg Island the Steacie Ice Cap. The NRC established the Steacie Memorial Fellowships and friends and colleagues, acting privately, set up the Steacie Memorial Fund to award the Steacie Prize. Both are awarded annually to encourage young scientists in the pursuit of excellence in science and to keep alive the memory of Steacie's personal convictions. York University would name their science library for Steacie. And in the nation's capital, Carleton University, founded by H.M. Tory, and where Steacie had smoothed many a meeting of the board of governors, chose to dedicate their new and splendid chemistry building to Steacie's memory:

> EMINENT SCIENTIST, TEACHER
> PRESIDENT OF THE NATIONAL
> RESEARCH COUNCIL OF CANADA
> AND CHAIRMAN OF THE BOARD OF
> GOVERNORS OF THIS UNIVERSITY

This citation would have pleased Steacie greatly. In his tireless crusade for the well-being of universities he had chaired many a meeting, but of his diverse roles this was the actual order by which he would have wished to be remembered.[17]

13

Epilogue

Steacie was an eminently confident man: confident in the importance of science and its achievements, in the institutional aspects of the National Research Council, and in the role each of these would play in the nation's future. This confidence had not been easily won, although an inherent streak of optimism helped. The certainty was a reward for one who had by discipline, fate, and innate ability conquered the times through which he had lived and its history – the loss of a father in the First World War, the devastation of the great depression, the spectacular rise of science and technology in the wake of the second world conflict, and the reinterpretation of Canada's place in the post-war hierarchy. His elevation to a position of influence within the Council would coincide with this last phase.

Steacie's generation was not the first to yearn for a show of national excellence. Being situated next to a powerful neighbour to the south provided a perpetual impetus, or as Rutherford so bluntly put it, on his appointment to the chair of physics at McGill, 'I am expected to do a lot of original work and to form a research school in order to knock the shine out of the Yankees.'[1] The fact that he did not tarry long enough to do so merely exacerbated the desire of those who followed to succeed, at least in the first set of goals. Even with this less ambitious target, the battle would be a protracted one.

The NRC possessed from the beginning all the complexities of the human condition. Governed by a code of procedure acceptable to the times, it faced daily decisions while carrying out its long-term objectives. It endeavoured to satisfy a desire for independence but was compelled to heed the conditions of external constraints. It consistently measured itself as to its successes or failures, and for much of its existence, the scale had tipped heavily toward the successes. Steacie's decade of leadership followed on from an era strongly marked by a unity of purpose between the Council and successive governments in fulfilling national goals. The institution he inherited was in a mood to take on some of the characteristics of the man on whom the responsibilities of leadership came to rest. Steacie delivered a highly personal style of leadership built upon a diversity of administrative and scientific abilities, accompanied by a relentless determination; with these came a generous and warm personality, whose strength and charm were compelling, and could easily win confidence and dispel dissent; it would evoke great admiration and even greater affection. Few who came into contact with him failed to recognize this rare combination and fewer still would remain unaffected. This did not always mean positively; a quick wit could often wound. But this was the same spirit which took habitual note of the needs of others, not infrequently at personal cost.[2]

Steacie's era was a special time in the development of modern science, that great hope of post-war society, a period of exuberant expansion, when even pessimists could see no limit on the horizon. This relatively simple era would be followed irrevocably by an eminently complex period. The public's vision of science and the role of scientists, for example, the trusting admiration enjoyed by Steacie's generation, would soon be jostled by hostile suspicion and demands for accountability. The mechanisms of communication at the scientific and political level which were so much a part of Steacie's day – largely a matter of personal interaction – would fragment with increasing rapidity. When Steacie was elected president of ICSU it was possible to ascribe to him the philosophy of 'the ascendancy of political considerations as long as the integrity of the scientific process' had not been compromised.[3] By 1971 others were advocating that scientists of political skill should learn to exploit relevant social beliefs to win a positive image for science and thus public support.[4] In actual fact, both administrators of science and governments would become increasingly manipulative, with the more vocal elements of society as umpire. For later generations of scientists and political leaders alike, the reality of compromise would become a necessity. Indeed, it was this particular political-scientific relationship which would become most highly fragmented. Afterward it was difficult to put the pieces together again; often, the values of the day had altered so drastically that the perception would depend upon which of the fragments came most easily to hand.

The Glassco Report on government organization was published in 1963, less than a year after Steacie's death. In addition to its broad view of government organization, it included brief reviews of several 'special areas of administration,' one of which was the NRC. From its assessment would develop a post-Steacie image of the NRC from which, remaining uncontradicted, there would be no turning back. One of the original purposes of the government in devoting money to research, the report pointed out, was to encourage and stimulate Canadian industry. 'From being a primary goal this has, over the years, been relegated to being little more than a minor distraction.' This particular piece of criticism was not aimed specifically at the NRC, but a few assuredly were. The Council had been established to promote research in industry and in universities. The first of these objectives, in the commission's considered opinion, had not been fulfilled. 'For this,' the commissioners proffered, 'the rather academic orientation of the National Research Council and its preoccupation with basic research may be in part responsible.'[5] This view would come to dominate with growing intensity the decades of debate ahead.

A few years later a Senate committee chaired by Senator Lamontagne began hearings on science policy; the first volume of their findings was delivered in 1970. Its aftermath, following on the Glassco Report, would evolve into controversies whose ardour has yet to abate.[6]

The Lamontagne Report introduced Steacie's decade of leadership thus:

Dr. Steacie was a skilled and noted researcher in fundamental science, who spent considerable effort to preserve the integrity and freedom of fundamental science and took effective steps to promote its development. He was concerned about the possible conflict between science and technology and worried that science's importance might be diminished by the glamour and utility of technology.[7]

This was certainly an unfortunate choice of words to describe Steacie's philosophy on science and technology which are interdependent, rarely in conflict, and never, in Steacie's opinion, glamorous. But it was another statement which would create, in the end, a breach between the NRC and government hitherto unknown. In its day, the report noted, the NRC had assumed many functions, a manifestly difficult task for one Council and one administrative unit. 'As a distinguished pure scientist, Dr. Steacie solved this difficulty by assuming in practice that NRC was mainly a university laboratory of basic research.'[8] This was an astonishing conclusion. It ignored the work of NRC's applied and engineering divisions, and overlooked information which Steacie himself had given to a parliamentary committee in 1960. 'In our own [NRC] lab, although our basic function is as an applied lab, we are doing something in the neighborhood of 20% of our effort in pure science.'[9] This simple statement had little impact on the views which the Lamontagne committee would maintain, for the received message had long ago been recorded and would be replayed on many successive occasions. Instead, the widespread feeling to which the Glassco Report referred, that fundamental research was the only activity adequately recognized within the National Research Council, would persist, amplified and perpetuated in the Lamontagne Report, and repeatedly appealed to by critics ever since.

The shortcomings of the Lamontagne committee's report as an objective account of the functioning and leadership of the NRC resulted, at least in part, from their less than thorough search for information. The committee did not once visit the NRC during its long proceedings.[10] Its lack of interest was presumably explained in a curious statement made by Senator Lamontagne when replying to criticisms of his report:

It is not true to say that the Committee tried to give the impression that we were presenting a fair and objective statement of Canadian science ... According to our terms of reference, our purpose was to study science policy in its broadest sense, not Canadian science.[11]

Therein would rest a fatal omission. For the development of science in Canada was bound irrevocably to the institutional history of the National Research Council, a symbiosis which was uniquely Canadian. Steacie's policy decisions were taken in order to counter the enormous demand for trained scientists in Canada and to stem the flow, if possible, of Canadian scientists to the United States, by providing a strong national basis for research. In criticizing the policies and philosophies of Steacie's leadership, the report neglected to weigh adequately the historical nature of the relationship between science and government up to and including Steacie's era; the shortage of indigenous scientific personnel due to the lack of training facilities in universities;[12] or the pattern of Canada's industrial development vis-à-vis both Britain and the United States. All these were facts, all pointed to the youthfulness of a history which is strictly Canadian.

Nevertheless, the misunderstandings, such as they were, evolved from a process in which Steacie himself had played a leading role, albeit inadvertently. He was a persistent advocate of 'pure' science, a term almost guaranteed to confuse those unconcerned with the nature of science itself. And he believed that the progress of technology was inevitable; it was society which had to adjust. But the calm, well-informed society en masse, free and capable of making sensible, deliberate choices to which Steacie had so often addressed his remarks in the 1950s, was rarely to be found. The golden image of science and technology would rapidly tarnish with time and become a subject of increasing suspicion. Far from being its master, as visualized by Steacie, society would increasingly feel its victim.

In restrospect, the demands for change constituted another chapter of history whose time had come, the aftermath of the spectacular rise of science. Its predictable fall was accompanied by a simultaneous and unanticipated eruption of sociological and political studies of scientific norms, paradigms, and policies. This flurry of activity was not confined to Canada. In Europe and the United States, enthusiasts of science policy burned the midnight oil to examine and re-examine their nation's science, from whence it had come and whither it was going. Only in Canada did the criticisms eventually settle on a single institution and the man who had personified its existence for a decade.

In the final analysis, there is no simple or sovereign remedy to the complex tissues of dependence binding science, society, and government in modern times. Both science and technology are the results of progress in the pursuit of knowledge. As to how this knowledge can best serve society, there are as many answers as points of view. The radical changes in policies on science and technology over the past quarter-century have not fundamentally resolved problems such as the country's industrial needs or the optimum value to be expended on research in the scientific and technological sectors. Steacie perceived, as a primary goal, the need to build a strong science foundation in a nation where very little previously existed; all other requisites – such as more qualified personnel and industrial expansion – ought, in all reason, to be thus fulfilled. Given the social and political circumstances of the decade over which he presided, it is difficult to see why Steacie would have elected to proceed from any other premise. His insistence on a dichotomy, both literally and philosophically, between the 'pure' and 'applied' domains of science was strictly personal, in many ways puzzling, in retrospect unwise, and really not necessary. But of one thing we may be reasonably certain. The nation for which Steacie laboured so relentlessly to achieve scientific excellence, whether it is labelled 'pure' or 'applied,' would not have been grateful if he had opted for mediocrity. Even such a potential critic as the Glassco Commission could state: 'The Council ... has built for itself an enviable reputation for the excellence of its work and the high quality of its professional personnel.'[13] Certainly, industrial organizations such as the Mellon Institute agreed.[14] These accolades alone were no mean achievements. Changes would quite rightly be made, even as times demanded, but the wisdom of first apportioning blame in order to reconstruct has yet to be demonstrated.

The great man of an age, according to Hegel, is the one capable of putting into words the will of his age, tells his age what its will is, and accomplishes it. To his supporters, Steacie's wise and far-sighted actions have created a modern scientific tradition in Canada; according to his critics, by sheer strength of will and a persistent eloquence, he was able to convince a whole generation of government officials, politicians, and intellectuals that his vision of science was also that of the nation. Steacie's place in history is thus assured.

APPENDIX

E.W.R. Steacie's Scientific Publications

BOOKS AND MONOGRAPHS

1929 (With W.H. Hatcher and N.N. Evans) *Laboratory Exercises in General Chemistry*. Montreal: Renouf Publishing Co. pp. v+70
1931 (With O. Maass) *An Introduction to the Principles of Physical Chemistry*. New York: John Wiley and Sons. pp. viii+272
1939 (With O. Maass) *An Introduction to the Principles of Physical Chemistry*. Second edition. New York: John Wiley and Sons. pp. ix+395
1946 *Atomic and Free Radical Reactions*. A.C.S. Monograph No. 102. New York: Reinhold Publishing Co. pp. vii+548
1946 *Free Radical Mechanisms*. New York: Reinhold Publishing Co. pp. x+251
1954 *Atomic and Free Radical Reactions*. Second edition. A.C.S. Monograph No. 125. New York: Reinhold Publishing Co. pp. x+901

SCIENTIFIC PAPERS

1925 (With F.M.G. Johnson) The viscosities of the liquid halogens. *J. Amer. Chem. Soc.* **47**, 754
1926 (With F.M.G. Johnson) The solubility and rate of solution of oxygen in silver. *Proc. Roy. Soc.* A, **112**, 542
1928 (With F.M.G. Johnson) The solubility of hydrogen in silver. *Proc. Roy. Soc.* A, **117**, 662
1929 (With F.J. Toole) Single crystals of silver. *J. Amer. Chem. Soc.* **51**, 1134
1929 (With O. Maass) An attempt to determine the osmotic pressures of very dilute solutions. *Trans Roy. Soc. Can.* III, **23**, 203
1930 The kinetics of the heterogeneous thermal decomposition of methyl formate. *Proc. Roy. Soc.* A, **127**, 314

1930 A convenient form of gas combustion pipet. *J. Amer. Chem. Soc.* **52**, 2811
1930 The rate of coagulation of silver hydrosol. *J. Phys. Chem.* **34**, 1848
1930 (With H.N. Campbell) The thermal decomposition of ethyl ether on the surface of platinum. *Proc. Roy. Soc.* A, **128**, 451
1930 (With G.B. Graham) The solubility of water vapour in solid inorganic compounds at high temperatures. *J. Phys. Chem.* **34**, 2098
1931 The thermal decomposition of diazomethane. *J. Phys. Chem.* **35**, 1493
1931 (With Richard Morton) The thermal decomposition of gaseous propionaldehyde on the surface of platinum. *Canad. J. Res.* **4**, 582
1931 Solubility as a complicating factor in adsorption measurements at gas-solid interfaces. *J. Phys. Chem.* **35**, 2112
1931 (With H.A. Reeve) A modified flow method for measuring the velocities of gas reactions. *Canad. J. Res.* **5**, 448
1931 (With W.H. Hatcher and F. Howland) The oxidation of acetaldehyde. *Canad. J. Res.* **5**, 648
1932 The kinetics of the oxidation of gaseous acetone. *Canad. J. Res.* **6**, 265
1932 The specific nature of energy exchange in unimolecular reactions. *J. Amer. Chem. Soc.* **54**, 1695
1932 Energy exchange in unimolecular reactions. I. The decomposition of mixtures of dimethyl and diethyl ether. *J. Phys. Chem.* **36**, 1562
1932 (With H.A. Reeve) Energy transfer between complex gas molecules and solid surfaces. *Trans. Roy. Soc. Can.* III, **26**, 75
1932 The homogeneous unimolecular decomposition of mixtures of gaseous aliphatic ethers. *Trans. Roy. Soc. Can.* III, **26**, 103
1932 Solubility and activated adsorption. *Trans. Faraday Soc.* **28**, 617
1932 (With W.H. Hatcher and Frances Howland) The kinetics of the oxidation of gaseous acetaldehyde. *Canad. J. Res.* **7**, 149
1932 (With H.A. Reeve) The decomposition of dimethyl ether on the surface of platinum. *J. Phys. Chem.* **36**, 3074
1933 Cross-activation in the unimolecular decomposition of mixtures of gaseous methyl and ethyl ether. *J. Chem. Phys.* **1**, 313
1933 The kinetics of the thermal decomposition of methyl ethyl ether. *J. Chem. Phys.* **1**, 618
1933 (With J.S. Tapp and O. Maass) An investigation of the density of a vapor in equilibrium with a liquid near the critical temperature. *Canad. J. Res.* **9**, 217
1933 (With E.M. Elkin) A comparison of the catalytic activity of liquid and solid surfaces. The decomposition of methanol on solid and liquid zinc. *Proc. Roy. Soc.* A, **142**, 457
1934 (With C.F.B. Stevens) The effect of a magnetic field on the linear rate of crystallization. *Canad. J. Res.* **10**, 483

1934 (With R.D. McDonald) The thermal decomposition of gaseous methyl iodide. *Canad. J. Res.* **10**, 591

1934 (With G.T. Shaw) The homogeneous unimolecular decomposition of gaseous alkyl nitrites. II. The decomposition of ethyl nitrite. *J. Chem. Phys.* **2**, 345

1934 (With E.M. Elkin) Further investigations on catalysis by liquid metals. *Canad. J. Res.* **11**, 47

1934 (With A.C. Plewes) The kinetics of the oxidation of mixtures of ethylene and acetaldehyde. *Proc. Roy. Soc.* A, **146**, 72

1934 (With Ernest Solomon) The kinetics of the homogeneous thermal decomposition of ethyl ether at pressures up to two hundred atmospheres. *J. Chem. Phys.* **2**, 503

1934 Die enzymatische Rohrzuckerinversion in schwerem Wasser. *Z. phys. Chem.* B, **27**, 6

1934 (With G.T. Shaw) The homogeneous unimolecular decomposition of gaseous methyl nitrite. *Proc. Roy. Soc.* A, **146**, 388

1934 (With H.V. Stovel) The rate of adsorption of ethylene by silica and nickel. *J. Chem. Phys.* **2**, 581

1934 (With J.W. McCubbin) The decomposition of nitrous oxide on the surface of platinum. I. The retarding effect of oxygen. *J. Chem. Phys.* **2**, 585

1934 (With A.C. Plewes) The kinetics of the oxidation of gaseous hydrocarbons. *Canad. Chem. Metall.* **18**, 216

1934 (With A.C. Plewes) The kinetics of the oxidation of gaseous hydrocarbons. II. The oxidation of ethane. *Proc. Roy. Soc.* A, **146**, 583

1934 (With R.D. McDonald) The kinetics of the reaction between gaseous methyl alcohol and nitrous oxide. *J. Phys. Chem.* **38**, 1031

1934 (With W.H. Hatcher and S. Rosenberg) The kinetics of the oxidation of gaseous propionaldehyde. *J. Phys. Chem.* **38**, 1189

1935 The oxidation of ethane. *Chem. & Ind.* **54**, 62

1935 Die enzymatische Spaltung von Salicin durch Emulsin in schwerem Wasser. *Z. phys. Chem.* B, **28**, 236

1935 (With R.D. McDonald) The kinetics of the thermal decomposition of gaseous methyl iodide. *J. Amer. Chem. Soc.* **57**, 488

1935 (With R.D. McDonald) Nitrous oxide as an oxidising agent in the gaseous state. *Canad. J. Res.* **12**, 711

1935 (With W.H. Hatcher and J.F. Horwood) The kinetics of the decomposition of gaseous glyoxal. *J. Chem. Phys.* **3**, 291

1935 (With G.T. Shaw) The homogeneous unimolecular decomposition of gaseous alkyl nitrites. III. The decomposition of n-propyl nitrite. *J. Chem. Phys.* **3**, 344

1935 (With K.H. Geib) Austauschreaktionen mit D-Atomen. *Z. phys. Chem.* B, **29**, 215

1935 (With W.H. Hatcher and J.F. Horwood) The kinetics of the oxidation of gaseous glyoxal. *J. Chem. Phys.* **3**, 551

1935 (With G.T. Shaw) The homogeneous unimolecular decomposition of gaseous alkyl nitrites. IV. The decomposition of iso-propyl nitrite. *Proc. Roy. Soc.* A, **151**, 685

1935 (With K.H. Geib) Exchange reactions involving atomic deuterium. *Trans. Roy. Soc. Can.* III, **29**, 91

1936 (With R.D. McDonald) The kinetics of the oxidation of gaseous hydrocarbons. III. The oxidation of acetylene. *J. Chem. Phys.* **4**, 75

1936 (With D.S. Calder) The homogeneous unimolecular decomposition of gaseous alkyl nitrites. V. The decomposition of methyl nitrite at low pressures. *J. Chem. Phys.* **4**, 96

1936 (With W.H. Hatcher and S. Rosenberg) The kinetics of the decomposition of ethyl ether at high pressures. *J. Chem. Phys.* **4**, 220

1936 (With S. Rosenberg) The decomposition of methyl nitrite at high pressures. *J. Chem. Phys.* **4**, 223

1936 (With J.W. McCubbin) The decomposition of nitrous oxide on the surface of platinum. II. The effect of foreign gases. *Canad. J. Res.* B, **14**, 84

1936 (With E.M. Elkin) Catalysis by fusions: A reply to the paper by Adadurow and Didenko. *J. Amer. Chem. Soc.* **58**, 691

1936 (With N.W.F. Phillips) The reactions of deuterium atoms with methane and ethane. *J. Chem. Phys.* **4**, 461

1936 (With W.McF. Smith) The homogeneous unimolecular decomposition of gaseous alkyl nitrites. VI. The decomposition of n-butyl nitrite. *J. Chem. Phys.* **4**, 504

1936 (With S. Katz, S.L. Rosenberg and W.McF. Smith) The homogeneous unimolecular decomposition of gaseous alkyl nitrites. VII. A general discussion of the effect of chemical configuration on the reaction rate. *Canad. J. Res.* B, **14**, 268

1937 (With S. Katz) The homogeneous unimolecular decomposition of gaseous alkyl nitrites. VIII. The decomposition of ethyl and n-propyl nitrites at low pressures together with a general discussion of the results for the entire series. *J. Chem. Phys.* **5**, 125

1937 (With N.W.F. Phillips) The thermal decomposition of ethane. A note on the paper by H. Sachsse. *J. Chem. Phys.* **5**, 275

1937 (With W.A. Alexander) The use of deutero compounds as indicators for the presence of free radicals in organic decomposition reactions. *J. Chem. Phys.* **5**, 372

1937 (With H.O. Folkins) The decomposition of nitrous oxide on a silver catalyst. *Canad. J. Res.* B, **15**, 237

1937 The reaction of deuterium atoms with methane at high temperatures. *Canad. J. Res.* B, **15**, 264

1937 (With W.A. Alexander) Free radicals in organic decomposition reactions. I. The thermal decomposition of mixtures of methyl ether and deutero-acetone. *Canad. J. Res.* B, **15**, 295

1938 The kinetics of the reaction $H + C_2H_6 = CH_4 + CH_3$. *J. Chem. Phys.* **6**, 37

1938 (With W.McF. Smith) A new method for the analysis of mixtures of chlorine, phosgene, and nitrosyl chloride. *Canad. J. Res.* B, **16**, 1

1938 (With W.McF. Smith) The unimolecular decomposition of gaseous chloropicrin. *J. Chem. Phys.* **6**, 145

1938 (With N.W.F. Phillips) The mercury photosensitized decomposition of ethane. *J. Chem. Phys.* **6**, 179

1938 (With I.E. Puddington) The kinetics of the decomposition reactions of the lower paraffins. I. *n*-butane. *Canad. J. Res.* B, **16**, 176

1938 The kinetics of the decompositions of the lower paraffins. *Canad. Chem. and Process Ind.* **22**, 325

1938 The kinetics of elementary reactions of the simple hydrocarbons. *Chem., Rev.* **22**, 311

1938 (With N.A.D. Parlee) The reaction of oxygen atoms with methane. *Canad. J. Res.* B, **16**, 203

1938 (With G. Shane) The polymerization of isobutene. *Canad. J. Res.* B, **16**, 210

1938 (With N.W.F. Phillips) A source of mercury resonance radiation of high intensity for photochemical purposes. *Canad. J. Res.* B, **16**, 219

1938 (With W.McF. Smith) The kinetics of the decomposition of chloropicrin at low pressures. *Canad. J. Res.* B, **16**, 222

1938 (With I.E. Puddington) The kinetics of the decomposition reactions of the lower paraffins. II. Isobutane. *Canad. J. Res.* B, **16**, 260

1938 (With N.W.F. Phillips) The mercury photosensitized decomposition of ethane. II. The production of hydrogen and the mechanism of the reaction. *Canad. J. Res.* B, **16**, 303

1938 (With W.A. Alexander and N.W.F. Phillips) The mercury photosensitized decomposition of ethane. III. The reaction in the presence of added deuterium. *Canad. J. Res.* B, **16**, 314

1938 (With Roger Potvin) A source of cadmium resonance radiation of high intensity and some preliminary results on the cadmium photosensitized reaction of hydrogen and ethylene. *Canad. J. Res.* B, **16**, 337

1938 (With I.E. Puddington) The kinetics of the decomposition reactions of the lower paraffins. III. Propane. *Canad. J. Res.* B, **16**, 411

1939 (With H.O. Folkins) The kinetics of the decomposition reactions of the lower paraffins. IV. The role of free radicals in the decomposition of n-butane. *Canad. J. Res.* B, **17**, 105

1939 (With N.A.D. Parlee) The reaction of hydrogen atoms with propane and the mechanism of the paraffin decompositions. *Trans. Faraday Soc.* **35**, 854

1939 (With Roger Potvin) The cadmium photosensitized reactions of ethane. *J. Chem. Phys.* **7**, 782

1939 (With N.A.D. Parlee) The reaction of hydrogen and deuterium atoms with propane. *Canad. J. Res.* B, **17**, 371

1940 (With H.O. Folkins) The kinetics of the thermal decompositions of the lower paraffins. V. The nitric oxide inhibited decomposition of n-butane. *Canad. J. Res.* B, **18**, 1

1940 The quenching of mercury resonance radiation by ethylene. *Canad. J. Res.* B, **18**, 44

1940 (With Roger Potvin) Cadmium photosensitized reactions of ethylene. *Canad. J. Res.* B, **18**, 47

1940 (With Gerald Shane) The kinetics of the decomposition reactions of the lower paraffins. VI. Ethane. *Canad. J. Res.* B, **18**, 203

1940 (With D.J. Dewar) Mercury photosensitized reactions of propane. *J. Chem. Phys.* **8**, 571

1940 (With E.A. Brown) The reaction of hydrogen atoms with butane. *J. Chem. Phys.* **8**, 734

1940 (With R.L. Cunningham) The mercury photosensitized decomposition of ethane. IV. The reaction at high temperatures. *J. Chem. Phys.* **8**, 800

1940 (With Gerald Shane) The kinetics of the decomposition reactions of the lower paraffins. VII. The nitric oxide inhibited decomposition of ethane. *Canad. J. Res.* B, **18**, 351

1941 (With D.J. LeRoy and R. Potvin) The cadmium photosensitized reactions of propane. *J. Chem. Phys.* **9**, 306

1941 Photosensitization experiments with various metal vapours. Contribution to the conference 'The Primary Process in Photochemistry' May 3–4, 1940. *Ann. N.Y. Acad. Sci.* **41**, 187

1941 (With D.J. LeRoy) The mercury photosensitized reactions of ethylene. *J. Chem. Phys.* **9**, 829

1942 (With D.J. LeRoy) The polymerization of ethylene photosensitized by 5^1P_1 cadmium atoms. *J. Chem. Phys.* **10**, 22

1942 (With H. Habeeb and D.J. LeRoy) Zinc photosensitized reactions of ethylene. *J. Chem. Phys.* **10**, 261

1942 (With D.J. LeRoy) The mercury photosensitized reactions of ethylene at high temperatures. *J. Chem. Phys.* **10**, 676

1942 (With D.J. LeRoy) The cadmium ($^3P_1$) photosensitized reactions of the lower olefins. *J. Chem. Phys.* **10**, 683

1942 (With D.J. LeRoy) Experimental methods of determining the activation energies of elementary reactions. *Chem. Rev.* **31**, 227

1943 (With D.J. LeRoy) The quenching of cadmium resonance radiation (3261Å) by hydrocarbons and other gases. *J. Chem. Phys.* **11**, 164

1944 (With D.J. LeRoy) The resonance emission of cadmium hydride bands in cadmium photosensitized reactions of hydrocarbons. *J. Chem. Phys.* **12**, 34

1944 (With D.J. LeRoy) Activation energies of elementary reactions. *Ann. Tab. phys. Const.* (1942, 1944), Section 602 (c), pp. 1–81

1944 (With D.J. LeRoy) The mercury photosensitized polymerization of acetylene. *J. Chem. Phys.* **12**, 117

1944 (With D.J. LeRoy) Determination of nitric oxide using solid reagents. *Industr. Engng. Chem. (Anal.)* **16**, 541

1944 (With D.J. LeRoy) The reaction of hydrogen atoms with acetylene. *J. Chem. Phys.* **12**, 369

1944 (With H.E. Gunning) The mercury photosensitized polymerization and hydrogenation of butadiene. *J. Chem. Phys.* **12**, 484

1945 (With G.M. Harris) The reaction of hydrogen atoms with acetone. *J. Chem. Phys.* **13**, 554

1945 (With G.M. Harris) The reaction of hydrogen atoms with dimethyl mercury. *J. Chem. Phys.* **13**, 559

1945 (With B. deB. Darwent) The mercury photosensitized reactions of propane at low pressures. *J. Chem. Phys.* **13**, 563

1946 (With H.E. Gunning) The mercury photosensitized reactions of propylene and isoprene. *J. Chem. Phys.* **14**, 57

1946 (With H.E. Gunning) The reaction of methyl radicals with butadiene and isoprene. *J. Chem. Phys.* **14**, 534

1946 (With H.E. Gunning) The mercury photo-sensitized reactions of isobutene. *J. Chem. Phys.* **14**, 544

1946 (With B. deB. Darwent and A.F. van Winckel) The mercury photosensitized reactions of diethyl ether. *J. Chem. Phys.* **14**, 551

1946 (With H.E. Gunning) The mercury photosensitized reactions of 1-butene and 2-butene. *J. Chem. Phys.* **14**, 581

1947 (With B. deB. Darwent and W.R. Trost) Elementary reactions involving the lower paraffins. *Disc. Faraday Soc.* **2**, 80

1947 Radiation chemistry and photochemistry. *Proc. Chemical Institute of Canada Conference on Nuclear Chemistry, McMaster University, Hamilton*, pp. 174–8

1948 (With M.K. Phibbs and B. deB. Darwent) The mercury photosensitized reactions of ethylene oxide. *J. Chem. Phys.* **16**, 39

1948 The relation of radiation chemistry to photochemistry. *J. Phys. and Coll. Chem.* **52**, 441

1948 (With B. deB. Darwent) The effect of variations in surface on the photolysis of acetone. *J. Chem. Phys.* **16**, 230

1948 (With W.R. Trost and B. deB. Darwent) The reaction of hydrogen atoms with some oxygen-containing organic compounds. *J. Chem. Phys.* **16**, 353

1948 (With W.R. Trost) The reaction of hydrogen atoms with the paraffin hydrocarbons. *J. Chem. Phys.* **16**, 361

1948 (With B. deB. Darwent) The mercury photosensitized reactions of ethane. *J. Chem. Phys.* **16**, 381

1948 (With B. deB. Darwent) The purification of neopentane by mercury photosensitization. *J. Amer. Chem. Soc.* **70**, 2285

1948 Photosensitized reactions of hydrocarbons. *Research, Lond.* **1**, 541

1948 (With H.E. Gunning) A further investigation of the mercury photosensitized polymerization of propylene. *J. Chem. Phys.* **16**, 926

1948 Photosensitized reactions. *Canad. J. Res.* B, **26**, 609

1948 (With R.A. Marcus and B. deB. Darwent) The mercury photosensitized reaction of dimethyl ether. *J. Chem. Phys.* **16**, 987

1948 Photosensitization. *Trans. Roy. Soc. Can.* III, **42**, 1

1949 Bond strengths in simple hydrocarbons. *Chem. Can.* **1**, No. 1, 33

1949 (With H.E. Gunning) The mercury photosensitized polymerization of cyclopropane. *J. Chem. Phys.* **17**, 351

1949 (With A. Cambron) Chemistry Division of the National Research Council of Canada. *Research, Lond.* **2**, 225

1949 (With B. deB. Darwent) The mercury photosensitized reactions of neopentane. *Canad. J. Res.* B, **27**, 181

1949 (With R.A. Marcus) Der sterische Faktor bei der Rekombination von Methyl-Radikalen. *Z. Naturf.* **4a**, 332

1950 (With C.R. Masson) Primary step in the mercury photo-sensitized decomposition of propane and of hydrogen. *J. Chem. Phys.* **18**, 210

1950 (With A.F. Trotman-Dickenson) The reactions of methyl radicals with hydrocarbons. *J. Amer. Chem. Soc.* **72**, 2310

1950 (With A.F. Trotman-Dickenson) The reactions of methyl radicals. I. The photolysis of acetone. *J. Chem. Phys.* **18**, 1097

1951 (With H.I. Schiff) The reactions of H and D atoms with cyclic and paraffin hydrocarbons. *Canad. J. Chem.* **29**, 1

1951 (With E.J. Caule) The photolysis of mercury dipropyl. *Canad. J. Chem.* **29**, 103

1951 (With D.M. Miller) The recombination of methyl radicals. *J. Chem. Phys.* **19**, 73

1951 (With E.J.Y. Scott) The mercury photosensitized reactions of benzene at high temperatures. *Canad. J. Chem.* **29**, 233

1951 (With A.F. Trotman-Dickenson and J.R. Birchard) The reactions of methyl radicals. II. The abstraction of hydrogen atoms from paraffins. *J. Chem. Phys.* **19**, 163

1951 (With A.F. Trotman-Dickenson) The reactions of methyl radicals. III. The abstraction of hydrogen atoms from olefins. *J. Chem. Phys.* **19**, 169

1951 (With S. Bywater) The mercury ($^3P_1$) sensitized decomposition of normal and isobutane at high temperatures. *J. Chem. Phys.* **19**, 172

1951 (With J.G. Calvert) Vapor phase photolysis of formaldehyde at wavelength 3130Å. *J. Chem. Phys.* **19**, 176

1951 (With C.R. Masson) The mercury photo-sensitized decomposition of methyl chloride. *J. Chem. Phys.* **19**, 183

1951 Review of 'Chemical Kinetics' by K.J. Laidler. *J. Amer. Chem. Soc.* **73**, 1900

1951 (With S. Bywater) The mercury ($^3P_1$) photo-sensitized decomposition of propane at high temperatures. *J. Chem. Phys.* **19**, 319

1951 (With S. Bywater) The mercury ($^3P_1$) photo-sensitized reaction of ethane at high temperatures. *J. Chem. Phys.* **19**, 326

1951 (With A.F. Trotman-Dickenson) The reactions of methyl radicals. IV. The abstraction of hydrogen atoms from cyclic hydrocarbons, butynes, amines, alcohols, ethers and ammonia. *J. Chem. Phys.* **19**, 329

1951 (With A.F. Trotman-Dickenson) The reactions of methyl radicals. *J. Phys. and Coll. Chem.* **55**, 908

1951 (With K.J. Ivin) The disproportionation and combination of ethyl radicals: the photolysis of mercury diethyl. *Proc. Roy. Soc.* A, **208**, 25

1951 (With M. Szwarc) Note on the temperature-independent factors of elementary reactions. *J. Chem. Phys.* **19**, 1309

1951 The decomposition of organic compounds. *Chem. Can.* **3**, No. 12, 29

1951 (With M.H.J. Wijnen) Photolysis of 2,2',4,4'-tetradeuterodiethyl ketone. *Canad. J. Chem.* **29**, 1092

1952 (With K.O. Kutschke and M.H.J. Wijnen) Mechanism of the photolysis of diethyl ketone. *J. Amer. Chem. Soc.* **74**, 714

1952 (With T.G. Majury) The reaction of CH_3 and CD_3 radicals with hydrogen and deuterium. *J. Chem. Phys.* **30**, 197

1952 (With M.H.J. Wijnen) The reaction of ethyl radicals with deuterium. *J. Chem. Phys.* **20**, 205

1952 (With F.A. Raal) The reaction of methyl radicals with some halogenated methanes. *J. Chem. Phys.* **20**, 578

1952 (With R.W. Durham) The reaction of methyl radicals with nitric oxide, and the rate of combination of methyl radicals. *J. Chem. Phys.* **20**, 582

1952 (With T.G. Majury) The reactions of CH_3 and CD_3 radicals with hydrogen and deuterium. *Canad. J. Chem.* **30**, 800

1952 (With K.J. Ivin and M.H.J. Wijnen) Reactions of ethyl radicals. *J. Phys. Chem.* **56**, 967

1953 (With R.J. Cvetanovic) Photolysis of acetone-hydrogen chloride mixtures. *Canad. J. Chem.* **31**, 158

1953 (With R.J. Cvetanovic and F.A. Raal) Photolysis of mixtures of acetone and some chlorinated methanes. *Canad. J. Chem.* **31**, 171

1953 (With L. Breitman) The use of infrared spectrophotometry for the estimation of small quantities of some halogenated hydrocarbons. *Canad. J. Chem.* **31**, 328

1953 (With R.W. Durham) The photolysis of azoisopropane. *Canad. J. Chem.* **31**, 377

1953 (With M.H. Jones) The reaction of methyl radicals with isobutane. *Canad. J. Chem.* **31**, 505

1953 The reactivity of free radicals. General introduction. *Disc. Faraday Soc.* **14**, 9

1953 (With T.G. Majury) The reactions of methyl radicals with the hydrogen isotopes. *Disc. Faraday Soc.* **14**, 45

1953 (With Richard E. Rebbert) Photolysis of mercury dimethyl. *Canad. J. Chem.* **31**, 631

1953 (With E. Whittle) The reactions of methyl radicals with the hydrogen isotopes. *J. Chem. Phys.* **21**, 993

1953 (With M.H. Jones) The photochemical decomposition of azomethane. *J. Chem. Phys.* **21**, 1018

1953 (With V.A. Crawford) The thermal decomposition of *n*-butane. *Canad. J. Chem.* **31**, 937

1953 (With F.P. Lossing) Photochemistry. *Annu. Rev. phys. Chem.* **4**, 323

1953 (With Richard E. Rebbert) The photolysis of mercury dimethyl in the presence of hydrocarbons. *J. Chem. Phys.* **21**, 1723

1953 Present status of radical mechanisms for organic decompositions. *Chem. Can.* **5**, No. 12, 31

1954 (With Richard E. Rebbert) The photolysis of mercury dimethyl with 3-ethyl pentane. *Canad. J. Chem.* **32**, 40

1954 (With L. Mandelcorn) Methyl ethyl ketone in the photolysis of acetone vapor. *Canad. J. Chem.* **32**, 79

1954 (With Richard E. Rebbert) The photolysis of mercury dimethyl with deuterium. *Canad. J. Chem.* **32**, 113

1954 (With H. Blades and A.T. Blades) The kinetics of the pyrolysis of toluene. *Canad. J. Chem.* **32**, 298

1954 (With P. Ausloos) The photochemical decomposition of azoethane. *Bull. Soc. chim. Belg.* **63**, 87

1954 (With L. Mandelcorn) The photolysis of acetone above 300°C. *Canad. J. Chem.* **32**, 331

1954 (With D.H. Derbyshire) The photolysis of dimethyl mercury in hexane solution at low temperatures. *Canad. J. Chem.* **32**, 457

1954 (With L. Mandelcorn) Addition of methyl radicals to unsaturated hydrocarbons. *Canad. J. Chem.* **32**, 474

1954 (With R.E. Dodd) The combination of methyl radicals: photolysis of acetone at low pressures. *Proc. Roy. Soc.* A, **223**, 283

1954 (With P. Ausloos) The reactions of methyl and ethyl radicals with diethyl ketone. *Canad. J. Chem.* **32**, 593

1954 (With Arthur T. Blades) Some aspects of the toluene pyrolysis. *Canad. J. Chem.* **32**, 1142

1955 (With P. Ausloos) The reaction of methyl radicals with CH_3CHO and CH_3CDO. *Canad. J. Chem.* **33**, 31

1955 (With P. Ausloos) The photolysis of biacetyl. *Canad. J. Chem.* **33**, 39

1955 (With P. Ausloos) Some complicating factors in the photolysis of acetone. *Canad. J. Chem.* **33**, 47

1955 (With M.J. Ridge) The photolysis of acetone in the presence of hydrogen bromide. *Canad. J. Chem.* **33**, 383

1955 (With M.J. Ridge) The photolysis of mixtures of acetone and some halogenated hydrocarbons. *Canad. J. Chem.* **33**, 396

1955 (With H.G. Oswin and R. Rebbert) Photolysis of acetone in the presence of mercury dimethyl. *Canad. J. Chem.* **33**, 472

1955 (With P.B. Ayscough and J.C. Polyanyi) The vapor phase photolysis of hexafluoroacetone in the presence of methane and ethane. *Canad. J. Chem.* **33**, 743

1955 (With P. Ausloos) The photolysis of methyl ethyl ketone. *Canad. J. Chem.* **33**, 1062

1955 (With S. Bywater) Mechanisms for the thermal decomposition of hydrocarbons. In *The Chemistry of Petroleum Hydrocarbons*, edited by B.T. Brooks, Cecil E. Boord, S.S. Kurtz, and L. Schmerling, vol. **2**, pp. 1–25. New York: Reinhold Publishing Co.

1955 (With J.B. Farmer, F.P. Lossing, and D.G.H. Marsden) Mass spectrometric detection of radicals in the mercury photosensitized decomposition of acetone. *J. Chem. Phys.* **23**, 1169

1955 (With R. Pieck) The photolysis of acetone in the liquid phase: the gaseous products. *Canad. J. Chem.* **33**, 1304

1955 (With P. Ausloos) The vapor-phase photolysis of acetic acid. *Canad. J. Chem.* **33**, 1530
1955 (With R.K. Brinton) Photolysis of diethyl ketone at low pressures: the pressure dependency of the combination of ethyl radicals. *Canad. J. Chem.* **33**, 1840
1956 (With P.B. Ayscough) The reactions of trifluoromethyl radicals with propane, *n*-butane and isobutane. *Canad. J. Chem.* **34**, 103
1956 (With H. Gesser) The photolysis of ketene in the presence of hydrogen. *Canad. J. Chem.* **34**, 113
1956 (With P.B. Ayscough) The photolysis of hexafluoroacetone. *Proc. Roy. Soc.* A, **234**, 476
1956 Reactions of radicals in gaseous systems (The Liversidge Lecture). *J. Chem. Soc.* p. 3986
1957 (With B.C. Spall) The mechanism of the photolysis of acetamide. *Proc. Roy. Soc.* A, **239**, 1
1957 (With Ikuzo Tanaka) Sensitized photo-ionization in far ultraviolet. *J. Chem. Phys.* **26**, 715
1957 (With Ikuzo Tanaka) Sensitized photo-ionization. *Canad. J. Chem.* **35**, 821
1957 (With G.O. Pritchard) The gas phase reaction of methyl radicals with hexafluoroacetone. *Canad. J. Chem.* **35**, 1216
1958 (With Hideo Okabe) The fluorescence and its relationship to photolysis in hexafluoroacetone vapor. *Canad. J. Chem.* **36**, 137
1958 (With D.E. McElcheran and M.H.J. Wijnen) The photolysis of methyl cyanide at 1849Å. *Canad. J. Chem.* **36**, 321
1958 (With D.G.L. James) Reactions of the ethyl radical. I. Metathesis with unsaturated hydrocarbons. *Proc. Roy. Soc.* A, **244**, 289
1958 (With D.G.L. James) Reactions of the ethyl radical. II. Addition to unsaturated hydrocarbons. *Proc. Roy. Soc.* A, **244**, 297
1958 (With G.H. Miller and G.O. Pritchard) The photolysis of perfluoro di-*n*-propyl ketone. *Z. phys. Chem.* **15**, 262 (Bonhoeffer Memorial Issue)
1958 (With D.G.L. James) Reactions of the ethyl radical. III. The effect of deuteration upon the metathetical reaction. *Proc. Roy. Soc.* A, **245**, 470
1958 (With K.O. Kutschke) A reply to Long on the activation energy of CH_3+H_2. *J. Phys. Chem.* **62**, 866
1958 (With G. Giacometti) The gas phase reactions of perfluoro-*n*-propyl radicals with methane and ethane. *Canad. J. Chem.* **36**, 1493
1958 (With Glenn H. Miller) The reactions of perfluoro-*n*-propyl radicals with hydrogen and deuterium. *J. Amer. Chem. Soc.* **80**, 6486
1959 (With G. Giacometti and H. Okabe) The wavelength- and temperature-

dependence on the fluorescence efficiency and of the primary photochemical yield in hexafluoroacetone vapor. *Proc. Roy. Soc.* A, **250**, 287

1959 (With K.O. Kutschke) The chemistry of free radicals in the gas phase. Reprinted from *Vistas in Free Radical Chemistry*, edited by W.A. Waters, p. 162. London: Pergamon Press (Kharasch Memorial)

1959 The decomposition of organic compounds. *J. Indian Chem. Soc.* **36**, 362 (Lecture given at a meeting of the Indian Chemical Society in Calcutta, 9 January)

1959 (With R.J. Cvetanovic and H.E. Gunning) Primary step in the mercury photosensitized reactions of olefins. *J. Chem. Phys.* **31**, 573

1959 (With A. Bruce King) The photolysis of trifluoromethyl cyanide. *Canad. J. Chem.* **37**, 1737

1959 Decomposition of organic compounds (in Russian). *Vestn. Akad. Nauk S.S.S.R.* **8**, 18

1960 (With G. Giacometti, H. Okabe, and S.J. Price) The photolysis and the fluorescence of perfluoro diethyl ketone. *Canad. J. Chem.* **38**, 104

1960 (With P.J. Boddy) Hydrogen atom abstraction by ethyl-d_5 radicals. Part I. *Canad. J. Chem.* **38**, 1576

1960 Article on free radicals. *McGraw Hill Encyclopedia*, p. 496

1960 (With J.F. Henderson) The photolysis of deuterated acetone and of deuterated acetone-hydrogen mixtures. *Canad. J. Chem.* **38**, 2161

1961 (With P.J. Boddy) Hydrogen atom abstraction by ethyl-d_5 radicals. Part II. *Canad. J. Chem.* **39**, 13

Notes

APSP   Advisory Panel on Scientific Policy
BMF   *Biographical Memoirs of Fellows of the Royal Society* (London)
CISTI Canada Institute for Scientific and Technical Information
CJMD   C.J. Mackenzie Diaries, Mackenzie Papers, PAC
CJMP   C.J. Mackenzie Office Papers, NRC
CSC   Civil Service Commission
DBS   Dominion Bureau of Statistics
DRB   Defence Research Board
MA   McGill Archives, McGill University, Montreal
OH   Oral History Project, NRC
OMP   Otto Maass Papers, McGill Archives
PAC   Public Archives of Canada, Ottawa
SP   Steacie Papers, PAC

1. EARLY DAYS AND MCGILL

1 Leo Marion, speech at the dedication of the Steacie Building for Chemistry, Carleton University, 22 Oct. 1965. The account of Steacie's childhood is constructed from conversations with his family and acquaintances of former neighbours.
2 Richard Preston, *Canada's RMC: A History of the Royal Military College* (Toronto: University of Toronto Press 1969), chap. 2
3 Ibid., 225
4 Record Group 32, vol. 689, PAC
5 See Yaffe, 'History of Department of Chemistry'; J.T. Edward, 'McGill Chemistry: 1000th PH D Coming Up,' *Canadian Chemical News* (Apr. 1984), 12–17.

6 O. Maass to C.J. Mackenzie, 2 Mar. 1961, OMP
7 R.F. Ruttan to Mr Justice Greenshields, 1 Apr. 1918, OMP
8 Private communication, Dr Carol Maass
9 O. Maass to J.D. Griffith Davies, 5 Dec. 1946; O. Maass to F. Cyril James, 15 Dec. 1952, OMP
10 BMF, 10 (1964), 257–81
11 Yaffe, 'History of Department of Chemistry,' 21
12 *McGill News* (Summer 1941), 56
13 Steacie to O. Maass, 2 Oct. 1939, OMP
14 *McGill News* (Summer 1941), 56
15 E.W.R. Steacie, 'Science and the University: Does Government Help?' *Queen's Quarterly* LXI, no. 3 (Autumn 1954), 313–18
16 Private communication. To be fair, the university did attempt to rid itself of the unwelcome rodents. 'I wish to thank you sincerely for the fine job you did in clearing the biological building of its vermin infestation,' Principal E.A. Morgan wrote to Professor W.H. Hatcher of the chemistry department in 1936 (Yaffe, 'History of Department of Chemistry,' 1). But clearly it was only a temporary lull and the vermin returned happily to their abode.
17 'Single Crystals of Silver,' *Journal of American Chemical Society* 51 (1929), 1134–5
18 *CBC Times* 2–8 June 1957, 11
19 Frost, *McGill University*, II: 157
20 R.V.V. Nicholls Collection, MA
21 McGill University calendar
22 O. Maass and E.W.R. Steacie, *An Introduction to the Principles of Physical Chemistry* (New York: John Wiley 1939)

2. A TASTE OF EUROPE AND THE YEARS AFTER

1 The award was funded under the auspices of the Carnegie Foundation. *The Royal Society of Canada Proceedings and Transactions*, 1935, scholarships, xix.
2 See obituary of Bonhoeffer by Paul Harteck in *Journal of Colloid Science* 13 (1958), 1–2.
3 Todd, *A Time to Remember*, 19
4 Steacie to O. Maass, 2 Sept. 1934, OMP
5 Ibid., 7 Dec. 1934
6 Mendelssohn, *The World of Walther Nernst*, 144
7 Steacie to O. Maass, 7 Dec. 1934
8 Ibid.

9 Ibid.
10 Notes made by Mrs Steacie during 1934–35
11 *Obituary Notices of Fellows of the Royal Society*, 9 (London 1954), 3–13
12 Ibid.
13 Ibid.
14 Ibid.
15 On the Kapitza incident, see Hartcup and Allibone, *Cockroft and the Atom*, 74 et seq; Wilson, *Rutherford*, chap. 16, esp. 496, 523.
16 See E.H. Hiebert, 'Walther Nernst and the Application of Physics to Chemistry,' in *Springs of Scientific Creativity*, ed. Rutherford Aris, H.T. Davis, and R.H. Stuewer (Minneapolis: University of Minnesota Press 1983), 205.
17 *Montreal Gazette*, 3 Dec. 1935
18 J.M. Keynes, *The Economic Consequences of the Peace* (London: Macmillan 1919), 209, 210
19 BMF 9 (1963), 195
20 *Dictionary of Scientific Biography*, ed., C.C. Gillispie (New York: Charles Scribner's Sons 1970–), entry for Nernst by Erwin Hiebert. For some personal views on this period of German science, see Mendelssohn, *The World of Walther Nernst*; W. Jost, 'The First 45 Years of Physical Chemistry in Germany,' *Annual Review of Physical Chemistry* 17 (1966), 1–14; O. Frisch, *What Little I Remember* (Cambridge: Cambridge University Press 1979).
21 Fritz Haber (1868–1934), *Dictionary of Scientific Biography* 5 (1972), 623
22 W.A. Noyes Jr, 'A Victorian in the 20th Century' (Austin: University of Texas 1976)
23 Frost, *McGill University* 2: 190
24 Ibid., 195, 196, 197, 201, 202

3. A CHEMIST FINDS HIS CALLING

1 M. Gomberg, 'Radicals in Chemistry, Past and Present,' *Industrial and Engineering Chemistry* 20 (1928), 164
2 M. Gomberg, 'An Instance of Trivalent Carbon: Triphenylmethyl,' *Journal of the American Chemical Society* 22 (1900), 757–71; see also 'The Existence of Free Radicals,' *Journal of the American Chemical Society* 36 (1914), 1144–70; 'Organic Radicals,' *Chemical Reviews* 1 (1924), 91–141.
3 Quoted by A. Ihde 'The History of Free Radicals and Moses Gomberg's Contributions,' *Pure and Applied Chemistry*, 15 (1967), 1–13
4 On the failure of the radiation hypothesis in kinetics, see M.C. King and K.J. Laidler, *Archive for History of Exact Sciences* 30 (1984), 45–86.

5 M. Bodenstein and W. Dux, 'Photochemische Kinetik des Chlorknallgases,' *Zeitschrift für physikalische Chemie* 85 (1913) 297–328; ibid., 329–97
6 W. Nernst, 'Zur Anwendung des Einsteinschen Photochemischen Aquivalentgesetzes I,' *Zeitschrift für Elektrochemie* 24 (1918), 335–6
7 H.S. Taylor, 'Photosensitization and the Mechanism of Chemical Reaction,' *Transactions of the Faraday Society* 21 (1926), 560–8. It appears that the idea may not have originated with Taylor but from a number of contributions made by other chemists, including J.A. Christiansen and Hans Bäckstrom; see *Journal of Physical Chemistry* 28 (1924), 145–8; *Journal of the American Chemical Society* 49 (1927), 1460–72. I am indebted to F.H. Winslow for bringing this detail to my attention.
8 The frequently more elegant and relative ease of the physical approach to many a scientific problem was a perpetual presence in the scientific milieu and did not go unnoticed in the chemist's camp. Soddy, for one, noted with rancour 'chemistry has been termed by the physicists as the messy part of physics but that is no reason why the physicists should be permitted to make a mess of chemistry when they invade it'; quoted in *More Random Walks in Science*, ed. Robert L. Weber (Bristol and London: The Institute of Physics, 1982), 64.
9 For a complete list of Steacie's chemical publications see the Appendix of this book.
10 'The Oxidation of Ethane,' *Chemistry and Industry* 54, no. 3 (1935), 62–3
11 Personal communication
12 L.W. Douglas to O. Maass, 21 Mar. 1939, OMP
13 Memo, Whitby to McNaughton, 3 Dec. 1938, Record Group 77, vol. 297, L1-6-2-24, PAC
14 Eagleson to Steacie, 9 Mar. 1939; Steacie to Eagleson, 15 Mar. 1939; Maass to Eagleson 23 Mar. 1939, Record Group 32, vol. 689, PAC
15 Personal communication

4. THE MAKING OF AN INSTITUTION: THE NATIONAL RESEARCH COUNCIL OF CANADA

1 National Research Council of Canada handbook, c. 1948, 1
2 Melville, *Department of Scientific and Industrial Research* 24; Clark, *Sir Edward Appleton*, 108
3 Eggleston, *National Research in Canada*, 4; see also Thistle, *The Inner Ring*.
4 NRC, *Annual Report 1918, 1937*; Eggleston, *National Research in Canada*, 7
5 Proceedings and Transactions of the Royal Society of Canada, 1882, quoted in Corbett, *Henry Marshall Tory*, 157

6 Eggleston, *National Research in Canada*, 4
7 Honorary Advisory Council minutes, 21 Sept. 1917
8 NRC, *Annual Report, 1918*, 28
9 Hamor and Weidlein, *Glances at Industrial Research*, preface
10 Ibid., 21
11 B. Doern, 'The National Research Council: The Causes of Goal Displacement,' *Canadian Public Administration* 12, no. 2 (1970), 160–1
12 Eggleston, *National Research in Canada*, 15; Thistle, *The Inner Ring*, 97–8
13 Proceedings of the Special Committee Appointed to Consider the Matter of Development in Canada of Scientific Research, 1919, quoted in Doern, 'National Research Council,' 158
14 A.B. Macallum to L.E. Westman, 6 Dec. 1920, quoted in Thistle, *The Inner Ring*, 97–8
15 Corbett, *Henry Marshall Tory*, 158
16 Quoted in ibid., 169
17 Quoted in Eggleston, *National Research in Canada*, 39–40
18 National Research Council of Canada, *Annual Reports 1929–35*; Eggleston, *National Research in Canada*, 29–31; Corbett, *Henry Marshall Tory*, 165–8
19 Corbett, *Henry Marshall Tory*, 170
20 Program of the official opening, National Research Laboratories, Ottawa, 10 Aug. 1932
21 Corbett, Henry Marshall Tory, 179
22 The *Ottawa Citizen*, 23 Feb. 1935, reported that 'McNaughton is Mentioned for Important Post; Report in Parliamentary Circles is that Dr Tory Will Retire Within Few Months.' The post carried a salary of $15,000.
23 'Andrew George Latta McNaughton 1887–1966,' *Proceedings of the Royal Society of Canada* (1967), 5: 103
24 Swettenham, *McNaughton*, I: 323
25 Ibid., 324–5
26 NRC, *Annual Report, 1937*, 11
27 *McGill News*, Sept. 1934, 21, 36
28 Pickersgill, *Mackenzie King Record*, I: 38
29 King to P.J.A. Cardin, Minister of Public Works, 12 Nov. 1937, NRC 6-6-38 (vol. 1). The prime minister had found time to read Eddington's *Science and the Unseen World* as part of his interest in the mystery of the Infinite.
30 NRC, *Annual Report, 1918*, 8. Of this sum, only just over $50,000 was disbursed.
31 Research Council Act, 1924

5. PROGRESS THROUGH PROBLEMS

1 See *BMF* 31 (1985), 411–34.
2 Thistle, ed., *Mackenzie-McNaughton Letters*, 9
3 Swettenham, *McNaughton*, I: 341–2
4 C.J. Mackenzie, notes for first meeting, 26 Oct. 1939, PAC Manuscript Group 30, B122, vol. 1, file 2
5 Thistle, ed., *Mackenzie-McNaughton Letters*, xiv
6 Robert Newton interviewed by D. Phillipson, 1975, OH
7 Leo Marion interviewed by D. Phillipson, 1975, OH
8 NRC, Proceedings of the Review Committee, 22nd meeting, 13–14 Sept. 1939
9 Memo, C.J. Mackenzie to Steacie, 15 Aug. 1961, Mackenzie private papers
10 'Otto Maass,' address given to the Chemical Society of McGill on the occasion of the retirement of Dr Maass by E.W.R. Steacie, 1955
11 O. Maass to Lester B. Pearson, Apr. 1961; Pearson replied 11 Apr. 1961: 'I am not a neutralist, an isolationist, or a pacifist, nor will the Liberal Party be, as long as I am its leader,' OMP.
12 C.J. Mackenzie to Steacie, 15 Aug. 1961, Mackenzie private papers
13 Thistle, ed., *Mackenzie-McNaughton Letters*, xvi
14 Eggleston, *National Research in Canada*, 125
15 Mackenzie to McNaughton, 18 Jan. 1940 in Thistle, ed., *Mackenzie-McNaughton Letters*, 17
16 C.J. Mackenzie, 'The National Research Council at War,' *Canadian Banker* (Jan. 1940), 143
17 E. Appleton to Mackenzie, 16 Jan. 1940, NRC 3-25-1-24, vol. 1
18 Mackenzie to Appleton, 5 Feb. 1940, ibid.
19 Quoted in Thistle ed., *Mackenzie-McNaughton Letters*, 3, 4
20 Originally called the War Technical and Scientific Development Fund, it was renamed the Sir Frederick Banting Fund after the death of Banting in 1941; see F.T. Rosser, 'The History of the Sir Frederick Banting Fund,' National Research Council of Canada 1960.
21 Mackenzie to McNaugton, 11 Feb. 1941, in Thistle, ed., *Mackenzie-McNaughton Letters*, 62
22 Mackenzie to McNaughton, 2 Jan. 1943, in Thistle, ed., *Mackenzie-McNaughton Letters*, 126
23 Ibid., 121
24 CJMD, 10 June 1943
25 In 1930, F.E. Lathe was made director of the Division of Research Information, but this was dissolved by the end of the decade. He became attached to the chemistry division and continued to be paid his director's salary. See

also N.T. Gridgeman's 'History of the Division of Chemistry,' NRC archives
26 Mackenzie to McNaughton, 27 Oct. 1941, in Thistle, ed., *Mackenzie-McNaughton Letters*, 98
27 Memo from R. Newton, 28 May 1943, NRC 4-P8-2, vol. 1
28 F.E. Lathe, 22 Aug. 1943, NRC 4-P8-2, vol. 1
29 NRC, Proceedings of the Review Committee, 16–17 Sept. 1943
30 Mackenzie to McNaughton, 28 Sept. 1943, in Thistle, ed., *Mackenzie-McNaughton Letters*, 130
31 Ibid., 132
32 Bothwell and Kilbourn, *C.D. Howe*, 186

6. LESSONS IN SCIENTIFIC DIPLOMACY

1 See Robert Bothwell, *Eldorado: Canada's National Uranium Company* (Toronto: University of Toronto Press 1984)
2 N.T. Gridgeman, 'History of the Division of Chemistry,' NRC archives
3 Steacie to Maass, draft cable, NRC 17-13E-8, vol. 1
4 *New York Times*, 5 May 1940; *Ottawa Journal*, 6 May 1940
5 *Ottawa Citizen*, 8 May 1940
6 J.D. Cockcroft to R.W. Boyle, 28 May 1940, NRC C17-13E-8, vol. 2
7 In the spring of 1940, Cockcroft was one of the handful of British scientists who agreed that the government should investigate the possibility of making a uranium bomb; see Hartcup and Allibone, *Cockcroft and the Atom*, 119.
8 See memo from S.J. Cook to Major-General W. Elkins, 25 May 1940, NRC 17-13E-8, vol. 1.
9 Maass to Steacie, 1 June 1940; Steacie to Maass, 3 June 1940; Maass to Steacie, 7 June 1940, NRC 17-13E-8, vol. 1
10 Cypher No. 831, 11 June 1940, NRC 17-13E-8, vol. 1
11 Cypher No. 879, 17 June 1904, ibid.
12 NRC 17-13E-8, vol. 1, 19 June 1940
13 Ibid., vol. 2, 16 July 1940
14 On his return to the UK. Cockcroft helped raise $5,000 for Laurence's work via CIL, the Canadian counterpart of ICI. See also Hartcup and Allibone, *Cockcroft and the Atom*, 101.
15 But see French views of the state of affairs in Weart, *Scientists in Power*; P. Craig, 'Frédéric Joliot and France's Nuclear Heritage,' *New Scientist*, 7 Feb. 1985, 16–19.
16 CJMD, 25 Sept. 1942

17 Leo Yaffe interviewed by author.
18 J.L. Heilbron, R.W. Seidel, and B.R. Wheaton, 'Lawrence and His Laboratory: Nuclear Science at Berkeley,' *Lawrence Berkeley News Magazine* (Berkeley: University of California 1981), 42
19 For an account of the group's work, see Glenn Seaborg, *History of Met Lab., Section C-1,* vol. 1 (1942–43), vol. 2 (1943–44) (Berkeley, Calif.: Lawrence Berkeley Laboratories PUB 112).
20 CJMD, 22 Feb. 1943
21 Conant to Mackenzie, quoted in Eggleston, *Canada's Nuclear Story,* 66. '49' was the code name for element 94, plutonium.
22 Quoted in Weart, *Scientists in Power,* 198, 199
23 Leo Yaffe interviewed by author.
24 CJMD, 16 Mar 1943
25 Mackenzie to G. Laurence, 18 Mar. 1943; Laurence to Mackenzie, 24 Mar. 1943, Record Group 77, vol. 283, PAC. In fact, George Laurence was not at all happy, perhaps understandably, at having Steacie brought in over his head. It was, he confessed almost thirty years later, 'very discouraging and distressing to me.' G.C. Laurence interview with D. Phillipson, OH, 1976.
26 Leo Yaffe interviewed by author
27 Quoted in Weart, *Scientists in Power,* 201
28 A discussion of the development of the Montreal lab and the appointment of Sir John Cockcroft will be found in Robert Bothwell, *Nucleus: The History of Atomic Energy of Canada Limited* (Toronto: University of Toronto Press 1988), which appeared after research on this book had been completed.
29 Memo, Cockcroft to division heads, 30 Nov. 1944, NRC D4-H6-31, vol. 1
30 G.C. Laurence, J.D. Cockcroft, quoted in Eggleston, *Canada's Nuclear Story,* 125, 127
31 Cockcroft, quoted in ibid., 127–8. Almost twenty years later, after Steacie's death, Cockcroft would recall the same moment of decision: 'I will never forget sitting with him [Steacie] on the banks of the Ottawa River at Balmer Bay when we decided on the site.' Cockcroft to Mrs Steacie, 30 Aug. 1962, Mrs Steacie, private papers; CJMD, 19 Aug. 1944: 'We recommended Chalk River and Howe agreed, so the Chalk River site was firmly fixed.'
32 ZEEP stood for Zero Energy Experimental Pile.
33 Quoted in Hartcup and Allibone, *Cockcroft and the Atom,* 130
34 Memo to Dr Cambron by Steacie on 'Organization of Work in Ottawa for Montreal Lab.,' 15 Dec. 1944, NRC D4-H6-31, vol. 1
35 Steacie teletype 14 Mar. 1945, NRC D4-H6-31, vol. 1
36 Quoted in Hartcup and Allibone, *Cockcroft and the Atom,* 130

37 CJMD, 7 Dec. 1944
38 The salary offered by NRC 'would be upwards of $10,000.' As director of the British Atomic Energy Establishment at Harwell, Cockcroft received £2,000.
39 Steacie to President, 27 Sept. 1946, Record Group 32, vol. 689, PAC
40 Thistle, ed., *Mackenzie-McNaughton Letters*, 76
41 CJMD June, July 1945; C.J. Mackenzie interview with W. Cherwinsky, 1974, NRC transcripts, 27; Mrs Mercer, C.J. Mackenzie's daughter, interviewed by author
42 Leo Yaffe interviewed by author
43 Quoted in J. Phinney Baxter, 3rd, *Scientists against Time* (Boston: Little, Brown 1946), 420
44 Memo, C.J. Mackenzie to Gordon Churchill, written shortly after Steacie's death but not sent, CJMP, vol. 2
45 On McNaughton's role, see Swettenham, *McNaughton* 3, chap. 4.
46 I am grateful to Dr H. Thode for sending the following personal comment: 'On a trip to Montreal I learned of the security break. It seems that Alan Nunn May, a physicst, spent part of an afternoon with his colleague Jules Gueron, a chemist, who was extracting Pu from dissolved irradiated uranium rods. The extraction procedures were explained to him and the final location of the Pu was pointed out. The next day Gueron reported to Steacie that he had lost the Pu and said he was the worst chemist in the world and should be fired.'
47 Canada, Report of the Royal Commission to investigate the facts relating to and the circumstances surrounding the communication, by public officials and other persons in positions of trust, of secret and confidential information to agents of a foreign power (Ottawa: King's Printer 1946), 384, 385, 447
48 Ibid., 384, 385, 386
49 Yaffe, *History of Department of Chemistry*, 39, 40

7. APPRENTICE IN STATESMANSHIP

1 Thistle, ed., *Mackenzie-McNaughton Letters*, 136; the full report is in CJMP.
2 NRC, Proceedings of the Review Committee, 23 Nov. 1945, 2
3 NRC, Annual Report of the Review Committee, 1946–47, Appendix B
4 Ibid., B2
5 C.J. Mackenzie interview, 22 May 1967, NRC transcripts, 109–10, also interviews 1971, 1975, OH. Mackenzie remembers the university offering the position to Steacie was probably the University of Toronto.

6 Terroux to Steacie, 6 Feb. 1948; Steacie to Terroux, 10 Feb. 1948, SP, vol. 1
7 Mackenzie interviewed by D. Phillipson, OH transcripts, 1, 2
8 Steacie to Herzberg, 21 Jan. 1947, Herzberg private papers. For a personal account of Herzberg's work, see G. Herzberg, 'Molecular Spectroscopy: A Personal History,' *Annual Reviews of Physical Chemistry* 36 (1985), 1–30.
9 Mackenzie to Herzberg, 13 Feb. 1947
10 Middleton, *Physics at the National Research Council*, 138, 139
11 C.J. Mackenzie, quoted in ibid., 135
12 Lord Dainton, private communication, 1983
13 NRC, Proceedings of the Review Committee, 21, 22 June 1946
14 For more details of the post-doctoral fellowship scheme, see NRC *Research News*, 1948; J.B. Marshall, 'Postdoctorate Fellowships of the National Research Council,' *Chemistry in Canada* 6, no. 9 (1954), 34–6; A.W. Tickner, 'Post-doctorate Fellowships in the National Research Laboratories,' *Canadian Chemical News* 37, no. 9 (1985), 17–19; N.T. Gridgeman, 'History of the Division of Chemistry,' NRC archives.
15 Brearley to Mackenzie, 13 Apr. 1948, NRC B18-19-3, vol. 2. A total of nineteen hundred post-doctoral fellows were to pass through the NRC.
16 Leo Marion interviewed by D. Phillipson, 1975, OH
17 These and following impressions of the fellowship program are based upon conversations and correspondences between the author and former fellows.
18 NRC, Proceedings of the Review Committee, 5 Dec. 1949, Appendix B
19 Much of this work was done with H.E. Gunning and D.J. LeRoy, both of whom established strong groups in chemical kinetics and held senior posts in Canadian universities in the post-war period.
20 See 'Science, Society and the Individual,' CBC radio talk, 8 Aug. 1959.
21 CJMD, 10 Oct. 1947; see also letters of congratulations, SP, vol. 1, file 1.
22 A case in point was the appeal by the Royal Society for funds to preserve Down House, the ancestral home of Charles Darwin. Steacie told Martin, 'It is my general feeling that most Canadians are a little fed up with this since they feel that the upkeep of historic sites in Britain is basically a job for Britain and that there is no particular reason why Canada should join in' (Steacie to D. Martin, 12 Dec. 1956, SP, vol. 8, file 5).
23 H.S. Taylor to Steacie, 27 Sept. 1958, SP, vol. 8, file 8. Although Steacie's regard for the Royal Society remained perennially high, it did not stretch to that of another Canadian fellow. In 1957 J. Stuart Foster of McGill suggested to Steacie that a group of advisers to the government could consist of fellows of the Royal Society living in Canada. Steacie replied, 'I think you will admit that if Canada is not to regard herself as a permanent colony

nothing could be more objectionable to the Government than to allow the Royal Society of London to appoint Scientific advisers to the Canadian Government by the process of electing them.' J.S. Foster to Steacie, 11 Oct. 1957; Steacie to Foster, 12 Nov. 1957, SP, vol. 1, file 6.
24 Steacie to J.L. Simonsen, 3 Nov. 1953, SP, vol. 8, file 8
25 Steacie to K.F. Bonhoeffer, 8 July 1947, 7 Oct. 1947, Bibliothek und Archiv für Geschichte der Max-Planck-Gesellschaft, Berlin, Bonhoeffer papers
26 Gordon to Mackenzie, 7 July 1948, Record Group 77, vol. 283, PAC
27 Mackenzie interviewed by D. Phillipson, OH transcripts, p. 110
28 J.B. McClinton to Steacie, 18 June 1959, NRC 9050-1, vol. 7
29 CISTI collection: Historical, Dr. E.W.R. Steacie. Steacie was appointed vice-president 1 July 1950 at a salary of $10,000, which was increased to $11,000 five months later; Record Group 32, vol. 689, PAC
30 W.A. Noyes Jr to Steacie, 26 Feb. 1957, SP, vol. 8, file 10
31 F. Blacet to author, 1 Nov. 1984
32 'Where Is Physics Going?' Reprinted in J.D. Babbitt, ed., *Science in Canada: Selections from the Speeches of E.W.R. Steacie*, 60–4 (hereafter cited as Babbitt)
33 Private communication to author
34 NRC, Annual Report of the Review Committee, 1951–52, Appendix B
35 Leopold Infeld, *Why I Left Canada* (Montreal and London: McGill-Queen's University Press 1978), 46, 57–9
36 Steacie to Herzberg, 29 Jan. 1947, Herzberg private papers

8. LEADER IN AN AGE OF CERTAINTY

1 Both cables in Record Group 32, vol. 689, PAC
2 Steacie to O. Maass, 26 Feb. 1952, OMP
3 CJMD, 7 Feb. 1952
4 See recollections in J.H. Parkin, *Aeronautical Research in Canada, 1917–1957* (Ottawa: NRC 1983).
5 Eggleston, *National Research in Canada*, 351
6 Quoted in Bothwell and Kilbourn, *C.D. Howe*, 261
7 Rt. Hon. Vincent Massey, *Our Debt to the Future*, Royal Society of Canada, Symposium presented on the Seventy-fifth Anniversary, 1957, ed. E.G.D. Murray (Toronto: University of Toronto Press 1958), 8
8 This figure would increase enormously. By the mid-1960s the U.S. federal budget on R&D alone was greater than its total budget before Pearl Harbor. See Don K. Price, *The Scientific Estate* (Cambridge, Mass.: Harvard University Press 1965).
9 C.V. Kidd, 'The Growth of Science and the Distribution of Scientists among Nations,' reprinted in Nelson, ed., *Politics of Science*, 88

10 M. Barkway, 'A Three-way Scientist,' *Saturday Night*, 30 Aug. 1952
11 *Time* magazine, 14 Sept. 1953
12 Barkway, 'A Three-way Scientist'
13 'Science and the National Academy' (1955), Babbitt, 81–91
14 Steacie to W.B. Lewis, 6 Nov. 1958, NRC 9000-7-1, vol. 6
15 Peter Medawar, *Pluto's Republic* (Oxford: Oxford University Press 1982), 267
16 Todd, *A Time to Remember*, 50
17 See, for example, W.W. Goforth to Steacie, 26 Jan. 1955, SP, vol. 6, file 4.
18 Mackenzie interviewed by M.W. Thistle and W. Eggleston, 1967, OH transcript, 128
19 'Science, Society and the Individual,' talk on CBC radio, Aug. 1959

9. THE POLITICS OF SCIENCE

1 'Engineering and Technological Education,' speech to Association of Universities of the British Commonwealth, Aug. 1958
2 'Science, Society and the Individual,' Aug. 1959
3 'The Development of Industrial Research in Canada' (Apr. 1961), Babbitt, 161–9
4 DBS, *Survey of Scientific & Industrial Laboratories in Canada* (Ottawa 1941)
5 Federal Government Expenditures, Report by the Department of Reconstruction and Supply and NRC, NRC C9000-7-1
6 Quoted from interview of General McNaughton and C.J. Mackenzie by Mr Beveridge, CBC/NRC Project 1966, NRC transcript, 13, 15
7 B.A. Gingras, T.W. West, 'National Research Council of Canada: 50 years of Scholarships,' *Engineering Journal* (Nov. 1967), 15–18; see also Council minutes for 13 Sept. 1946 and NRC *Annual Report* for 1945–6 and 1949–50.
8 President's Papers, 1945, University of Toronto Archives
9 'The Development of Industrial Research in Canada' (Apr. 1961), Babbitt, 161–9
10 'The Task of the University Today' (June 1960), Babbitt, 28–35
11 'The Impact of Society on Science,' Purvis Memorial Lecture to Society of Chemical Industry, Nov. 1957
12 See Eggleston, *National Research in Canada*, 282
13 Steacie to J.J. Deutsch, 25 July 1955, NRC L1-10-1-37, vol. 1
14 Steacie to D.M. Watters, 18 Sept. 1956, ibid.
15 R.B. Bryce to D.M. Watters, 24 Sept. 1956, ibid.
16 Summary of discussion at a private dinner meeting in Montreal, 22 Feb. 1955 on 'Scientific Research and Canadian Industry,' organized by the

Canadian Council for Economic Studies, prepared by Gilbert Jackson and Associates, Toronto, copy in SP, vol. 5, file 4
17 Memo from W.W. Goforth to Steacie, SP, vol. 5, file 4
18 H.B. Style, comments, SP, vol. 5, file 4
19 'The Development of Industrial Research in Canada' (Apr. 1961), Babbitt, 161–9
20 'The Impact of Society on Science' (Nov. 1957), Babbitt, 95–104
21 G.H. Shrum to Cyrias Ouellet, 15 Jan. 1958. Present copy from private source. How wrong Shrum was in his judgment of Steacie can be gleaned from a memo written by Steacie intended for the chairman of PCCSIR, dated 13 Nov. 1957. It is not clear whether the document was ever sent. 'If, however, it is desired to make a real attack on the whole problem of making Canadian basic research second to none then the program of university support needs to branch out in other directions. In the first place $50,000,000 has recently been placed at the disposal of the Canada Council for capital grants in the humanities and social sciences on a 50–50 basis with the universities. In view of the very large expansion of scientific enrollment which is being anticipated, of the increasing demands for scientists, and of the great expense of buildings and equipment for science, it is suggested that a major effort to develop Canadian scientific research could be made by putting the sum of at least $50,000,000 in the hands of the National Research Council for capital grants in science on terms similar to those for the existing grants in the Humanities.' See SP, vol. 1, file 6.
22 NRC, Proceedings of the 199th Meeting of the Council, Ottawa, 15 Nov. 1957, 2
23 Steacie to G.H. Shrum, 30 Jan. 1958, SP, vol. 3, file 12
24 Ibid.
25 J.W.T. Spinks, 'Proposal for a Royal Commission,' undated; Steacie to Spinks 30 Jan. 1958, SP, vol. 3, file 12
26 R.B. Miller to Steacie 1 Feb. 1958, NRC 3-3-C-188, vol. 1; Steacie to Miller, 6 Feb. 1958, SP, vol. 3, file 12
27 H.G. Thode to Steacie, 5 Feb. 1958; Steacie to Thode, 10 Feb. 1958, NRC 3-3-C-188, vol. 1
28 L. Katz to Steacie, 19 Feb. 1958, 'Memo on Discussions Regarding Federal Support for Scientific Research in Canadian Universities,' NRC 3-3-C-188, vol. 1
29 Steacie to J.W.T. Spinks, 26 Feb. 1958, SP, vol. 3, file 12
30 Steacie to J.D. Babbitt, 15 July 1958, NRC 9000-7-1, vol. 5. The advisory panel is discussed in chapter 10.
31 Agenda, Privy Council Office, 15 Dec. 1958, NRC 9000-7, vol. 6

32 Minutes, meeting of PCCSIR, 22 Dec. 1958, NRC 9000-7, vol. 6

10. SCIENCE AND GOVERNMENT: THE HEENEY REPORT

1 Granatstein, *The Ottawa Men*, 1, 231
2 Interview of General McNaughton and Dean Mackenzie by Mr Beveridge, CBC/NRC Archives project 1966, NRC transcript, 23
3 C.J. Mackenzie to Henry Tizard, 9 Apr. 1945, NRC 1453-41, vol. 1
4 CJMD, 13 Oct. 1944
5 See minutes of APSP meetings, NRC 9000-7-1. The panel consisted initially of representatives of six departments and agencies engaged in research, plus the Treasury Board. The president of NRC was the chairman and the secretary was provided by the Privy Council Office.
6 R.B. Bryce to Steacie, 22 Jan. 1954; Steacie to Bryce, 25 Jan. 1954, NRC 9000-7-1, vol. 3
7 R.B. Bryce to Steacie, 2 July 1953, NRC 9000-7-1, vol. 3
8 Steacie to R.B. Bryce, 9 July 1953, NRC 9000-7-1, vol. 3
9 Memorandum for the APSP, 26 Nov. 1953, NRC 9000-7-1, vol. 3
10 R.B. Bryce to Steacie, 5 Feb. 1954, NRC 9000-7, vol. 2
11 The previous year Steacie had voiced some unease in a memo (dated 13 Nov. 1957) to the new chairman of PCCSIR. 'The general feeling in the past seemed to be that a meeting of the committee [the PCCSIR] could only be concerned with the domestic details of various departments' operations and that this would not be of much interest to the various Ministers or of much value. I feel, however, that the lack of an overall scientific policy has been a handicap.' See SP, vol. 1, file 6.
12 See Diaries, Arnold Heeney Papers, PAC.
13 A.D.P. Heeney, 'The Things That Are Caesar's,' commencement address, Kenyon College, Ohio, 13 June 1955
14 Ibid.
15 Heeney to Steacie, 22 Nov. 1957, NRC, C3-12-C4-9, vol. 1
16 National Research Council of Canada, Comments on the Report of the Civil Service Commission of Canada, *Personnel Administration in the Public Service*, Ottawa, Dec. 1958. The final draft of the NRC document was signed by both Steacie and Rosser and dated 20 Feb. 1959; SP, vol. 7, file 12.
17 Ibid.
18 Steacie to the Rt. Hon. John Diefenbaker, 18 Feb. 1959, NRC C3-12-C4-9, vol. 1
19 The head of one government department during this period recalled his own feelings: 'Steacie had freedom and funds which I didn't have.' Only

personal admiration for Steacie prevented serious antagonism. The pay scales in most exempt agencies were also higher (private communication).
20 Steacie to Bryce, 24 Feb. 1959, NRC C3-12-C4-9
21 J. Diefenbaker to Steacie, 25 Feb. 1959, ibid.
22 Gordon Churchill to Steacie, 25 Feb. 1959, ibid.
23 Steacie to Churchill, 2 March 1959, ibid.
24 A.D.P. Heeney to C.J. Mackenzie, 22 Feb. 1947, NRC C9000-7, vol. 1
25 C.J. Mackenzie to A.D.P. Heeney, 28 Feb. 1947, NRC C9000-7, vol. 1
26 G.R. Pearkes to Zimmerman, 17 Mar. 1959, NRC C3-12-C4-9
27 Churchill to Steacie, 17 Mar. 1959, ibid.
28 H. Williamson to F.T. Rosser, 25 Mar. 1959, ibid.
29 *Montreal Star*, editorial, 10, 17 Aug.; articles 15, 17 Aug.; *Ottawa Journal*, editorial, 7 Feb., 27 July; articles 22, 29 Sept. 1959; quote from *Montreal Star*, 17 Aug. 1959
30 Quoted by I. Norman Smith in 'Steacie's Scientists vs. Civil Service Commission,' *Ottawa Journal*, 22 Sept. 1959
31 Steacie's annotated copy of the Heeney Report, SP, vol. 7, file 12
32 'Science, Society and the Individual,' series of CBC talks given by Steacie during August 1959
33 S.H.S. Hughes to Steacie, 16 July 1959, NRC C3-12-C4-9
34 Steacie to S.H.S. Hughes, 17 July 1959, ibid.
35 S.H.S. Hughes to Steacie, 28 July 1959, ibid.
36 'Heeney Answers Steacie Re: Scientists and Commission,' *Ottawa Journal*, 29 Sept. 1959
37 'The Civil Service Commission and the Outside Agencies' (1959), Babbitt, 136. The speech was written by Steacie for presentation at the Oct. 1959 conference of the Institute of Public Administration but was withdrawn because of the changing circumstances.
38 *Ottawa Journal*, 29 Sept. 1959
39 Arnold Heeney, *The Things That Are Caesar's: The Memoirs of a Canadian Public Servant* (Toronto: University of Toronto Press 1972), 147, 148
40 A. Heeney to Mrs Steacie (personal papers), 29 Aug. 1962

11. YEARS OF FULFILMENT

1 SP, vol. 3, file 12
2 See NRC 9050-1, vols. 3, 4, 1954-55.
3 'The Future of Scientific Research,' speech to the Defence College (1950)
4 J.H. Sword to Steacie 3 June 1955, SP, vol. 5, file 5
5 'Implications of the Atomic Age' (Aug. 1955), Babbitt, 105-11

6 *Ottawa Citizen*, 16 Aug. 1955
7 See clippings, SP, vol. 5, file 5.
8 Steacie to M.T. Bloom, 25 Aug. 1955, SP, vol. 5, file 5
9 G.D. Weaver to Steacie, 13 Jan. 1953; Steacie to Weaver, undated; reply from AECB is dated 21 Jan. 1954. Weaver's letter was probably misdated (year was 1954), NRC 9050-1, vol. 3.
10 G.D. Weaver to Steacie, 18 Feb. 1955; Steacie to Weaver, 1 Mar. 1955, NRC 9050-1, vol. 3
11 Speech, 'Implications of the Atomic Age'; Steacie, 'Comments on Mr [Ritchie] Calder's Remarks Concerning the Literacy and Sociological Attainment of Scientists,' *Professional Public Service*, Oct. 1953, 3, 12
12 L. Pauling, *No More War* (New York: Dodd, Mead & Co. 1958), 232
13 The Mainau Declaration of Nobel Laureates, Lake Constance, 15 July 1955
14 'The Impact of Society on Science' (Nov. 1957), Babbitt, 95–104
15 Personal communication, 1984; see also NRC 9050-1, vol. 8.
16 See Nelson, ed., *The Politics of Science*, 407.
17 Ibid., 409
18 Steacie to Bertrand Russell, 18 Mar. 1958, SP, vol. 3 file 12; Bothwell and Kilbourn, *C.D. Howe*, 273
19 Nelson, ed., *The Politics of Science*, 408
20 Hartcup and Allibone, *Cockcroft and the Atom*, 273–6, 282–3
21 Einstein to C.J. Mackenzie, 30 Apr. 1947, Record Group 77, vol. 283, PAC
22 G.D. Weaver to Steacie, 18 Feb. 1955, NRC 9050-1, vol. 3
23 Steacie to W.A. Noyes, 20 Feb. 1957, SP, vol. 8, file 3
24 Hamor to Steacie, 17 Oct. 1957, SP, vol. 6, file 1
25 C.E. Rea to Steacie, 11 Feb. 1960, SP, vol. 6, file 5
26 Private communication to author
27 J.G. Kirkwood to Steacie, 29 Apr. 1957, SP, vol. 1, file 6; Steacie to Kirkwood, 1 May 1957, SP, vol. 5, file 8
28 For details, see BMF 10 (1964), 257–81
29 Elizabeth Loosley to Steacie, 11 May 1955, SP, vol. 5, file 5
30 'The Organization of Scientific Knowledge' (Feb. 1953)
31 The Under-Secretary of State for External Affairs to Steacie, 2 Oct. 1957, SP, vol. 6, file 2
32 Steacie to G. Malloch, 13 Sept. 1957, SP, vol. 6, file 2
33 O. Maass to Steacie, 24 June 1958, SP, vol. 6, file 2
34 Personal communication to author from H.G. Thode, 25 July 1985
35 On background to Kaptiza and the Cambridge episode, see Wilson, *Rutherford*, chap. 16; Hartcup and Allibone, *Cockcroft and the Atom*, 72–7.
36 No account of Steacie's visit to the Soviet Union is available. I am indebted to

Mrs Steacie, Dr Ian McTaggart Cowan, and Professor H.G. Thode, who accompanied Steacie, for their memories of that visit. The delegation also included R.F. Farquharson and B.G. Ballard. See also NRC, Proceedings of 205th meeting of the Council, 9 Nov. 1959.
37 Notes for talk, 'Some Impressions of Soviet Science and Its Organizations,' Dec. 1960, SP, vol. 10
38 Speech to the International Union of Geodesy and Geophysics (3 Sept. 1957), Babbitt, 173–9
39 'Science and International Affairs' (4 Nov. 1960), Babbitt, 180–9
40 C.V. Carter, 'The Distribution of Scientific Effort,' *Minerva* (Winter 1963), 172–81
41 The report, when completed in 1963, showed that in the period 1952–62 twice as many British scientists and engineers with PH Ds emigrated to United States as to Canada; see 'Emigration of Scientists from the U.K.,' Royal Society Report, 1963.
42 Joan Butt, 'Brain Drain,' *NPL News* 11, 21 Nov. 1967; quoted from R.L. Weber, ed., *More Random Walks in Science: An Anthology* (Bristol and London: The Institute of Physics, 1982)
43 'The Future of Scientific Research' (1950), speech to the Defence College
44 'Industrial Research in Canada,' June 1956 (draft). Steacie gave considerable thought as to how these attitudes could be altered. In 1957 he addressed a memo with comments on industrial research to the chairman of PCCSIR, although it is not clear whether this was actually sent.

> There is no question that because of our industry being largely controlled from abroad a comparatively small amount of industrial research is done in Canada. There are, however, major difficulties in expanding industrial research. One of the main problems is that Canadian markets are relatively limited and it is thus difficult to justify too much research on an economic basis. In spite of this, however, the situation is improving steadily and the change in the last ten years is a very marked one. It is felt, therefore, that although the total amount of industrial research may not be what might be desired the trend is an excellent one. It is not easy to do much, other than missionary work, to expand industrial research ... One possible way of encouraging such research is to have research done in industry at Government expense, i.e. to set up contracts for research within industry. This has the advantage of starting research in industries which may not have been doing it. It has the disadvantage of producing an attitude in industry that research is a luxury which some-one else should pay for. Certainly, this latter attitude has been fostered in the aeronautical industry to such an extent that it is almost impossible to per-

suade any aeronautical firm anywhere in the world to do any research at its own expense. It is, however, obvious that such contracts for research can only be placed by those research organizations which are consumers of research as well as performers. It is thus difficult to see how one could justify such a contract placed by the National Research Council except in most unusual circumstances.

See memo dated 13 Nov. 1957, SP, vol. 1, file 6.

45 'The Development of Industrial Research in Canada' (Apr. 1961), Babbitt, 161–9
46 'The Role of the Pure and Applied Sciences in National Development' (Sept. 1958), speech to the National Federation of Canadian University Students
47 Speech, 'Development of Industrial Research in Canada'
48 Steacie, Memorandum to PCCSIR, Sept. 1961, NRC 9000-7-1, vol. 7
49 For details of early stages of IRAP, see NRC 2800-3. The implications of IRAP in its historical and political context are discussed in D.J.C. Phillipson, 'The Steacie Myth and the Institutions of Industrial Research,' *HSTC Bulletin* 7, no. 3 (1983), 117–34.
50 'Submission of the Canadian Manufacturers' Association to the Government of Canada with Respect to Industrial Research and Development,' NRC 1453–40. Steacie subsequently sent a memo to all senior and administrative members of NRC. 'During recent months there has been a very welcome increase in interest in research on the part of industry. Unfortunately, a by-product of this has been a considerable amount of ill-informed criticism of the general type, that, "NRC is only interested in high-brow science, and does nothing to help industry." It is, I think, desirable to correct this impression, and to emphasize whenever possible what we actually do to help industry.' Steacie asked for information on work and help provided to industry during the previous year. See memo 15 Feb. 1962, SP, vol. 4, file 16.
51 Steacie to J.C. Whitelaw, 'Comments on CMA Brief,' Jan 1962, SP, vol. 6, file 8
52 J.C. Whitelaw to Steaie, 18 Apr. 1962, ibid. Steacie's address, 'Research in Canadian Industry,' was delivered on 5 June 1962.

12. FINAL DAYS AND UNFINISHED BUSINESS

1 NRC, 54th Annual Report, 1961–62: 16–17
2 Steacie to General G.R. Laclavère, 15 Jan. 1962, ICSU Archives, Paris
3 D.W. Bronk to Steacie, 9 Apr. 1962; Minutes of 1st Meeting of the Committee on Future Structure of ICSU, Paris, 26–27 May 1962, NRC 1459–39, vol. 6
4 N. Ramsey, quoted in convocation address by W.A. Noyes Jr, 22 Oct. 1965, Carleton University, Ottawa

5 From correspondence in SP, vol. 4, file 4
6 Steacie to F.S. Dainton, 14 Aug. 1962
7 Steacie to C. Reid, 20 June 1958, SP, vol. 3, file 11
8 Memorandum from R.E. McBurney to NRC Advisory Committee on Industrial Research, 18 July 1962, NRC 4-1-7-3
9 Steacie to G. Steele, 6 July 1962, NRC L1-10-2-64, vol. 2
10 Daily journal, SP, vol. 9; D C. Martin, 'Portrait: E.W.R. Steacie,' ICSU *Review* 4 (1962), 169–71
11 Steacie to Carol Maass, 8 Aug. 1961, SP, vol. 8, file 4
12 Steacie to J. van Mieghan, 25 Aug. 1962, SP, vol. 4, file 14
13 Personal communication, Mar. 1985
14 'E.W.R. Steacie,' obituary by G. Herzberg in *Physics in Canada* 18, no. 5 (1962), 42–3
15 C.J. Mackenzie, pencil notes, personal papers
16 See SP, vol. 4, file 14
17 See NRC, Proceedings of 223rd Meeting, 4 Nov. 1964; the official opening of the E.W.R. Steacie Building for Chemistry, Carleton University, 22 Oct. 1965. More recently the Chemical Institute of Canada has initiated the E.W.R. Steacie Award in Photochemistry (1984).

13. EPILOGUE

1 Quoted in Wilson, *Rutherford*, 171
2 This was repeatedly stressed by former colleagues to the author. Typically, Steacie bequeathed his personal library to universities in deprived parts of the world, NRC 9050-1, vol. 10.
3 J.D. Babbitt, 'Profile of Dr. E W.R. Steacie. The New President of ICSU,' *New Scientist* 5 (Oct. 1962)
4 See, for example, essays in Nelson, *The Politics of Science*, and Y. Ezrahi, 'The Political Resources of American Science,' *Science Studies* 1, no. 2 (Apr. 1971), 117–23.
5 Canada, *The Royal Commission on Government Organization* (Glassco Report), vol. 4 (Ottawa 1963), 230, 231
6 *A Science Policy for Canada: Report of the Senate Special Committee on Science Policy* (Lamontagne Report), vol. 1 (Ottawa 1970). For insights into the controversies, see: issues of *Science Forum* (Ottawa), 1968–1976; G.B. Doern, 'The National Research Council: The Causes of Goal Displacement,' *Canadian Public Administration* 12 (Summer 1970), 140–84; F.R. Hayes, *The Chaining of Prometheus: Evolution of a Power Structure for Canadian Science* (Toronto: University of Toronto Press 1973); more recently, L. Dandurand,

'The Nature of the Politicization of Basic Science in Canada: NRC's Role 1945–1976' (PH D thesis, University of Toronto 1982); D.J.C. Phillipson, 'The Steacie Myth and the Institutions of Industrial Research,' *HSTC Bulletin* 7, no. 3 (1983), 117–34.
7 Lamontagne Report, vol. 1, 67
8 Ibid., 67–8
9 Canada, House of Commons, Special Committee on Research, Minutes of Proceedings and Evidence, 9 June 1960, 92. See also pp. 20, 21, 22 (27 May, 2 June 1960). Information in the Glassco Report supports this distribution of effort. 'The Ottawa laboratories are devoted to investigation in the fields of pure and applied chemistry, pure and applied physics and applied biology.' In addition, 'four engineering divisions are concerned with aeronautics, radio and electrical engineering, building research, and mechanical engineering. Regional laboratories at Halifax and Saskatoon slant their investigations toward local resources and economic interests of the Atlantic and Prairie Provinces respectively.' The figures for operating costs quoted are pure (basic) research divisions, $2.7 million; the other divisions including regional, $18.8 million (pp. 212, 213).
10 The invitation for such a visit was issued by then NRC president W.G. Schneider during 1968–69 and repeated in 1976. See Senate of Canada, Proceedings of the Standing Committee on Science Policy, 26 Mar. 1976.
11 'Senator Lamontagne Replies to the Critics of the Senate Science Report,' *Science Forum* 4, no. 3 (June 1971), 12
12 The senator and critics in general of Steacie's avid policy to increase research in Canadian universities should note with interest some remarks made in 1959 by British scientist Hugh S. Taylor, the president of the Woodrow Wilson National Fellowship foundation. Some 42 per cent of Canada's top scientists and 50 per cent of scholars had gone southward from Canada to enrich the academic life of the United States, Taylor told his audience. 'The United States is grateful for this support,' but it was time, Taylor thought, for American foundations to repay some of the intellectual debt that the United States had incurred from Canada. Proceedings of the National Conference of Canadian Universities and Colleges, 1959, University of Saskatchewan, 24.
13 Glassco Report, 212
14 See Paul J. Flory, executive director of research, Mellon Institute, to Steacie 25 Oct. 1957, SP, vol. 6, file 1: 'It was a great privilege to me to have the opportunity to discuss our plans with you, particularly in view of the very considerable similarity between the objectives which we have set for ourselves and those which you have pursued so effectively in your institution.'

Bibliography

Babbitt, J.D., ed. *Science in Canada: Selections from the Speeches of E.W.R. Steacie*. Toronto: University of Toronto Press 1965
Baxter, James P. *Scientists against Time*. Boston: Little, Brown and Co. 1946
Bothwell, Robert, and William Kilbourn. *C.D. Howe: A Biography*. Toronto: McClelland and Stewart 1979
Brebner, J. Bartlet. *Canada: A Modern History*. Ann Arbor: University of Michigan Press 1970
Clark, Ronald W. *Tizard*. London: Methuen 1965
– *Sir Edward Appleton*. Oxford: Pergamon Press 1971
Cochrane, Rexmond C. *Measures for Progress: A History of the National Bureau of Standards*. U.S. Department of Commerce 1966
Cockroft, John, ed. *The Organization of Research Establishments*. Cambridge: Cambridge University Press 1965
Cook, W.H. *My Fifty Years with NRC, 1924–1974*. Ottawa: National Research Council of Canada 1977
Corbett, E.A. *Henry Marshall Tory: Beloved Canadian*. Toronto: Ryerson Press 1954
Creighton, Donald. *The Forked Road, Canada 1939–1947*. Toronto: McClelland and Stewart 1976
Doern, Bruce G. *Science and Politics in Canada*. Montreal: McGill-Queen's University Press 1972
Dupree, A. Hunter. *Science in the Federal Government*. New York: Harper and Rowe 1957
Eggleston, Wilfrid. *Scientists at War*. Oxford: Oxford University Press 1950
– *Canada's Nuclear Story*. Toronto: Clarke Irwin 1965
– *National Research in Canada: The NRC 1916–1966*. Toronto: Clarke Irwin 1978

Finlay, J.L., and D.N. Sprague. *The Structure of Canadian Universities*. Scarborough, Ont.: Prentice Hall 1984

Frost, Stanley B. *McGill University for the Advancement of Learning, Volume 1 (1801–1895); Volume 2 (1895–1971)*. Kingston and Montreal: McGill-Queen's University Press 1980, 1984

Goodspeed, D.J. *A History of the Defence Research Board*. Ottawa: Edmond Cloutier 1958

Gowing, Margaret M. *Britain and Atomic Energy 1939–1945*. London: Macmillan 1964

Granatstein, J.L. *The Ottawa Men: The Civil Service Mandarins 1935–1957*. Toronto: Oxford University Press 1982

Gridgeman, N.T. *Biological Sciences at the National Research Council of Canada: The Early Years to 1952*. Waterloo, Ont.: Wilfrid Laurier University Press 1979

Hailsham, Viscount. *Science and Politics*. London: Faber and Faber 1963

Harmor, W.A., and E.R. Weidlein. *Glances at Industrial Research*. New York: Reinhold 1936

Hartcup, Guy, and T.E. Allibone. *Cockcroft and the Atom*. Bristol: Adam Hilger 1984

Hayes, F. Ronald. *The Chaining of Prometheus: Evolution of a Power Structure for Canadian Science*. Toronto: University of Toronto Press 1973

Hewlett, Richard G., and Oscar E. Anderson, Jr. *The New World, 1939/1946: A History of the United States Atomic Energy Commission*. Vol. 1. University Park: Pennsylvania State University Press 1962

Lithwick, N.H. *Canada's Science Policy and the Economy*. Toronto: Methuen 1969

McCrensky, Edward. *Scientific Manpower in Europe – A Comparative Study of Scientific Manpower in the Public Service in Great Britain and Selected European Countries*. London: Pergamon Press 1958

McInnis, Edgar. *Canada: A Political and Social History*. Toronto: Holt, Rinehart and Winston 1982

MacLennan, Hugh, ed. *McGill, the Story of a University*. London: George Allen and Unwin 1960

Melville, Sir Harry. *The Department of Scientific and Industrial Research*. London: George Allen and Unwin 1962

Mendelssohn, K. *The World of Walther Nernst: The Rise and Fall of German Science 1864–1941*. Pittsburgh: University of Pittsburgh Press 1973

Middleton, W.E.K. *Physics at the National Research Council of Canada 1929–1952*. Waterloo, Ont.: Wilfrid Laurier University Press 1979

– *Mechanical Engineering at the National Research Council of Canada, 1929–1951*. Waterloo, Ont.: Wilfrid Laurier University Press 1984

- *Radar Development in Canada: The Radio Branch of the National Research Council of Canada, 1939–1946.* Waterloo, Ont.: Wilfrid Laurier University Press 1981
Nelson, William R., ed. *The Politics of Science.* New York: Oxford University Press 1968
Noyes, W. Albert Jr. 'Memoirs: A Victorian in the 20th Century.' Austin: University of Texas, Department of Chemistry 1976
OECD Reviews of National Science Policy – Canada. Paris 1969
OEEC. The Problem of Scientific and Technical Manpower in Western Europe, Canada and the United States. Paris (c. 1956–1957)
Orlans, Harold, ed. *Science Policy and the University.* Washington, DC: Brookings Institute 1968
Parkin, J.H. *Aeronautical Research in Canada, 1917–1957: The Memoirs of J.H. Parkin.* 2 vols. Ottawa: National Research Council of Canada 1983
Phillipson, D.J.C. *Associate Committees of the National Research Council of Canada, 1917–1975.* Ottawa: National Research Council of Canada 1983
Pickersgill, J.W. *The Mackenzie King Record.* Vols. 1–4. Toronto: University of Toronto Press 1960–1970
Polanyi, Michael. *Science, Faith and Society.* Chicago: University of Chicago Press 1964; first published 1946
– ed. *Science and Freedom.* Report of the Harrisburg Congress. Boston: Beacon Press 1955
Research in Canada, Planning for the Coming Years. Chemical Institute of Canada, Quebec City 1945; published by Imperial Oil Limited, 1946
Royal Commission on Canada's Economic Prospects. *Final Report.* Ottawa 1957
Royal Commission on Government Organization (Glassco Report). Vol. 4, *Special Areas of Administration,* 23: Scientific Research and Development. Ottawa 1963
Senate of Canada. *Special Committee on Science Policy Report* (Lamontagne Report), vol. 1. Ottawa 1970
Spiegel-Rosing, I., and Derek de Solla Price. *Science, Technology and Society.* Beverly Hills: Sage 1977
Spinks, John. *Two Blades of Grass: An Autobiography.* Saskatoon: Western Producer Prairie Books 1980
Swettenham, John. *McNaughton.* 3 vols. Toronto: Ryerson Press 1968–69
Thistle, M. *The Inner Ring.* Toronto: University of Toronto Press 1966
– ed. *The Mackenzie-McNaughton Wartime Letters.* Toronto: University of Toronto Press 1975
Todd, Alexander. *A Time to Remember.* Cambridge: Cambridge University Press 1983

Weart, Spencer R. *Scientists in Power.* Cambridge, Mass.: Harvard University Press 1979

Weeks, Mary E. *Discovery of the Elements,* 6th ed., published by *Journal of Chemical Education,* Easton, Pa., 1956

Wiesner, Jerome B. *Where Science and Politics Meet.* New York: McGraw-Hill 1965

Wilson, David. *Rutherford, Simple Genius.* Cambridge, Mass.: MIT Press 1983

Zuckermann, Sir Solly. *Beyond the Ivory Tower.* London: Weidenfeld and Nicolson 1970

The War Histories of the National Research Council. These war histories covering NRC's activities during the Second World War were produced by the National Research Council of Canada in mimeographed form in 1946 and 1947. They are as follows:

History of the Wartime Activities of the Division of Applied Biology
War History of the Division of Chemistry
War History of the Division of Mechanical Engineering
War History of the Division of Physics
The War History of the Radio Branch
War History of the Associate Committees of the National Research Council
History of the Associate Committee on Medical Research by G.H. Ettinger
History of the Associate Committee on Aviation Medical Research, 1939–1945
History of the Associate Committee on Naval Medical Research, 1941–1945
Medical Research and Development in the Canadian Army during World War II, 1942–1946

Index

Advisory Panel on Scientific Policy 145, 149, 184
Alberta, University of 49, 142
Allmand, A.J. 18, 25–6
American Chemical Society 173
Appleton, Sir Edward 58, 87
Atomic Energy Control Board 159, 169
Atomic Energy of Canada Limited 116

Babbitt, J.D. 145
Banting, Sir Frederick 10, 70
Bell Laboratories 54–5
Bennett, R.B. 52, 53, 54
Bessborough, Earl of 52
Birchard, E.R. 107, 125
Blacet, Francis 110
Bonhoeffer, K.F. 18, 19–23, 106
Bowen, E.J. 105
Boyer, Raymond 92
Boyle, R.W. 61, 78, 99
brain drain: from Britain 180–1; from Canada 234 n.12
British Commonwealth Scientific Conference 116
Bronk, Detlev W. 188

Bryce, R.B. 136, 145, 149–51, 157
Bush, Vannevar 71, 148

California, University of 110, 193
Cambron, Adrien 87, 95
Canadian Association of Physicists 110, 145
Canadian Commentator 192
Canadian Council for Economic Studies 137, 181
Canadian Institute on Public Affairs 167–8
Canadian Manufacturers' Association 45, 184–5
Carleton University 125, 194
Chadwick, Sir James 71, 86
Chemical Institute of Canada 95, 109
chemical kinetics: early work on 34–7. *See also* Steacie, E.W.R.
chemistry division 62–3, 73, 94–6
Churchill, Gordon 145, 158–9, 192
Civil Service Commission 48, 151, 161
Cockcroft, J.D. 78, 79, 86–9, 90, 172
Compton, A.H. 71, 86
Conant, J.B. 71, 82
Cornell University 173
Cronyn Committee 48

Cvetanovic, R.J. 192

Dainton, F.S. 189
Dawson, Sir William 46
Defence Research Board 159, 184
Densmore, K.D. 102, 111
Deutsch, J.J. 135
Diefenbaker, John G. 156–7
Duncan, R.K. 47–8

Eagleson, S.P. 41, 71
Eaton, Cyrus 171–2
Eldorado Gold Mines Ltd. 77
Ensell, George 102
External Affairs, Department of 149, 176

Faraday Society 111, 127, 174
Financial Post 190
Flood, E.A. 70, 71
Frankfurt, University of 19

Geib, K.H. 21, 106
General Electric Company 54–5
Glassco Report 197, 198, 200, 234 n.9
Gomberg, Moses 34
Gordon, A.R. 107
Gouzenko revelations 91–2
Gray, J. Lorne 183
Guggenheim, E. 105

Halban, Hans 80–6
Hamor, W.A. 55
Heeney, A.D.P. 149, 151, 152–3, 162–3
Heeney Report, controversy over 153–61
Herzberg, Gerhard 98–9, 106, 114, 192
Hill, A.V. 69

Hinshelwood, Sir Cyril 105, 177
Holmes, Everett (cousin of EWRS) 5
Honorary Advisory Council on Scientific and Industrial Research 45, 124
Howe, C.D. 70, 71, 87, 122, 147, 148, 172; as chairman of PCCSIR 94, 101, 108, 117, 118, 149
Hughes, S.H.S. 162

Industrial Research Assistance Program 183–4, 190
industrial research in Canada: extent of 46, 55, 56–7, 131; Steacie's views on 137–9, 181–2
Infeld, Leopold 114
International Council of Scientific Unions 181, 187, 188
International Geophysical Year 139
International Union of Pure and Applied Chemistry 173, 188

Johnson, F.M.G. 10, 15

Kaiser Wilhelm Institute 54
Kapitza, Peter 26–7, 177–8
Katz, Leon 143
Keys, D.A. 97, 107
King, William Lyon Mackenzie 57, 58, 148
King's College (London) 24–5
Kowarski, Lew 87

Laclavère, G.R. 191–2
Lamontagne Report 197–8
Langmuir, Irving 71
Lathe, F.E. 72, 74
Laurence, G.C. 78, 80
Lewis, W.B. 88, 125

Index 241

Maass, Otto: early life and McGill 7–9, 11, 16–17, 41; wartime activities 64–6, 70, 73, 78–9; impact of Gouzenko revelations on 92; relations with Steacie 9, 30, 42–3, 177, 191
Macallum, A.B. 46, 49 156
Mackenzie, C.J. x, 9, 53, 61, 193; acting president 60; wartime responsibilities 61–2, 64, 66–73, 75–6; atomic energy project 79–83, 85, 88–90; president of NRC 94; post-war responsibilities 96–100, 107–8, 115, 116–18; relations with the government 147–8, 158–9; president of AECL 116; continuing relationship with NRC 118, 127, 183, 186, 191
Marion, Leo xi, 63, 73, 101, 119, 191
Maritime Regional Laboratory 127
Martin, Sir David 105, 181
Massey, Vincent 79
Massey Commission 149
McGill University 5, 7, 24, 32–3, 41, 92, 133, 189
McKim, F.L.W. 157
McNaughton, A.G.L. 48, 53–9, 60, 62, 91
Meighen, Arthur 51
Mellon Institute 47, 54–5, 173
Melville, Sir Harry 26
Montreal, University of 81
Montreal Gazette 29
Montreal laboratory 80–3
Montreal Star 160
Morrison, J.A. 111

National Academy of Sciences (U.S.) 174

National Bureau of Standards 48, 54, 150, 189
National Conference of Canadian Universities 129
National Physical Laboratory 54
National Research Council of Canada: origin and early years 45–9; under H.M. Tory 49–53; opening of the laboratories 52; under A.G.L. McNaughton 53–9; wartime expansion 69, 71; post-war planning 72, 94; Review Committee of 61, 63, 73–4, 94, 95, 100, 103, 113
NATO, Scientific Committee of 189
New York Times 192
Newton, Robert 61, 74, 89
Noyes, W.A. Jr. 31, 110, 173
nuclear research, NRC involvement in 80–8
nuclear weapons controversy 166–71

OECD 187, 189
Ottawa, University of 125
Ottawa Board of Trade 182
Ottawa Citizen 168
Ottawa Journal 156, 160, 168

Paneth, F.A. 81, 83, 87
Parkin, J.H. 61, 117, 118
Pauling, Linus 170–1
Pearkes, G.R. 159
Pearson, Lester B. 65
post-doctoral fellowships 100–3
Prairie Regional Laboratory 127
Privy Council Committee on Scientific and Industrial Research 45, 145, 151, 158
Puddington, I.E. 119, 122
Pugwash conferences 171–2

Rice, F.O. 38
Rideal, Sir Eric 105, 193
Robertson, Norman 117
Rockefeller Foundation 133
Rosser, F.T. 154–5
Royal Institution (London) 173
Royal Military College 5, 6
Royal Montreal Regiment 15
Royal Society (London) 69, 176–7, 180, 189
Royal Society of Canada 18, 109
Russell, Bertrand 171–2
Rutherford, Ernest 7, 26–7, 133, 195
Ruttan, R.F. 8

St Laurent, Louis S. 101, 166, 187
Santa Claus fund 70; *see also* War Technical and Scientific Development Fund
Saskatchewan Wheat Pool 122
Saturday Night 122
Saunders, Charles 48–9
Scientific and Industrial Research, Department of (U.K.) 45, 49, 68, 160
Scully, V.W.T. 73
Semenov, Nicolai 177
Shrum, G.H. 141, 143
Smart, General C.A. (uncle of EWRS) 5
Soddy, Frederick 7
Soviet Academy of Sciences 176–7
Spinks, J.W.T. 83, 98
Sputnik, the effects of 139–40
Steacie, Alice Kate (mother of EWRS) 3, 4
Steacie, Diana Jeannette (daughter of EWRS) 16
Steacie, Dorothy Catalina (wife of EWRS) 15

Steacie, Edgar William Richard: birth 3; boyhood 3–6; bachelor degree 9; doctoral degree 11; Sterry Hunt fellowship 11; appointed lecturer at McGill 11; attitude to religion 13; early married life 15–16; collaboration with Otto Maass 16; to Europe on fellowship 18; in Germany 18–24; in Britain 24–8; impressions of Germany 23–4, 29–30; research on chemical kinetics 37–41, 104; recruited by NRC 41–3; initiation as director of chemistry 61–3; wartime committee responsibilities 64; views on post-war research 74–5; attachment to Montreal lab 83–6; involvement in nuclear research 86–8; new approaches to post-war research 94–5; elected fellow of the Royal Society 104–5; president, Chemical Institute of Canada 109, 174; appointed vice-president of NRC 107–8; vice-president of IUPAC 174; appointed president of NRC 116; Baker Visiting lecturer (Cornell University) 111; president, Royal Society of Canada 174; Liversidge lecturer (Royal Institution) 173; views on industrial research 137–9, 181–2; views on university research 133; foreign associate, National Academy of Sciences (U.S.) 174; election to Soviet Academy of Sciences 176; visit to USSR 177–8; response to the Heeney Report 154–61; president, Faraday Society 174; illness 186–7; president of ICSU 187; death 192; tributes to 192–3; memorials to 194, 233 n.17

Steacie, John Richard Brian (son of EWRS) 16
Steacie, Richard (father of EWRS) 3, 4, 5
Stedman, D.F. 77

Taylor, H.S. 36, 105–6, 147–8, 234 n.12
Terroux, F.R. 97
Thode, H.G. 83, 143, 191, 192
Thompson, H.W. 105
Time 122
Times (London) 192–3
Tizard, Sir Henry 68, 69, 148
Todd, Alexander 19, 20, 160
Tolman, R.C. 71
Toole, F.J. 12–13
Toronto, University of 111, 114, 132
Toronto *Globe* 51
Tory, H.M. 45, 49–53, 57–8, 133, 156
Treasury Board 100, 135–6, 141, 165, 183–4, 190–1

UNESCO 189
university research support 109, 124, 132–7, 140–6, 186
uranium 78–9

Wallaceburg Rotary Club 175
War Technical and Scientific Development Fund 220 n.20
Watters, D.M. 136
Weaver, George D. 169, 173
Weidlein, E.R. 55
wheat rust 51–2
Wheaton, Ethel 193
Whitby, G.S. 41, 77
Whitelaw, J.C. 185
Wilgress, Dana 160
Winkler, Carl 92

York University 189, 194

Zimmerman, Hartley 7, 159, 183